BARBARA HENDERSON is an Inverness-based writer and drama teacher and the current Forth Bridge Writer-in-Residence. With her children's books, her energetic school visits take her across the length and breadth of Scotland, and sometimes beyond. *Scottish by Inclination* was Barbara's first foray into adult non-fiction and *Made from Girders* is her first collaboration with photographer Alan McCredie. Now that her three children have flown the nest, she shares her home with her long-suffering husband and a scruffy Schnauzer called Merry.

ALAN McCREDIE has been a professional photographer and filmmaker for over a decade, working with most major agencies in Scotland and beyond. He has specialised in theatre and television but is perhaps best known for his documentary and travel photography. A member of Documenting Britain photo collective, he is a Perthshire man lost to Leith.

THE FORTH BRIDGE, completed in 1890, and still the world's longest cantilever bridge, links Edinburgh to Fife and onwards to the North East and the Highlands, carrying around 200 trains a day. It was the world's first major steel structure and its famous cantilever design is recognised all over the world. In July 2015, the Forth Bridge became the sixth UNESCO World Heritage Site in Scotland.

By Barbara Henderson:

Fir for Luck, Cranachan Publishing, 2016
Punch, Cranachan Publishing, 2017
Wilderness Wars, Cranachan Publishing, 2018
Black Water, Cranachan Publishing, 2019
The Siege of Caerlaverock, Cranachan Publishing, 2020
The Chessmen Thief, Cranachan Publishing, 2021
Scottish by Inclination, Luath Press, 2021
The Reluctant Rebel, Luath Press, 2022
Rivet Boy, Cranachan Publishing, 2023

By Alan McCredie:

100 Weeks of Scotland, Luath Press, 2014
This is Scotland, Luath Press, 2014 with Daniel Gray
Scotland the Dreich, Luath Press, 2016
Scotland the Braw, Luath Press, 2019
Tribes of Glasgow, Luath Press, 2019 with Stephen Millar
Edinburgh the Dreich, Luath Press, 2021
'Mon the Workers, Luath Press, 2022 with Daniel Gray

Made from Girders

Our Forth Bridge

BARBARA HENDERSON
and
ALAN McCREDIE

Luath Press Limited
EDINBURGH
www.luath.co.uk

First published 2023

ISBN: 978-1-80425-104-1

The authors' right to be identified as authors of this book under the Copyright, Designs and Patents Act 1988 has been asserted.

The paper used in this book is recyclable. It is made from low chlorine pulps produced in a low energy, low emission manner from renewable forests.

Printed and bound by
Robertson Printers, Forfar

Typeset in 12 point Sabon LT
by Main Point Books, Edinburgh

Text © Barbara Henderson
Images © Alan McCredie

*To Frank Hay, Karen Stewart and Miles Oglethorpe,
the three cantilevers who carried this book from shore to shore.
Thank you for your encouragement and endless practical help.*

Barbara Henderson

Contents

AUTHOR'S NOTE: BARBARA HENDERSON	9
PHOTOGRAPHER'S NOTE: ALAN McCREDIE	15
ARTIST Gordon Muir	19
BLUE BADGE TOUR GUIDE Andrew Simpson	25
CONSTRUCTION SUPERINTENDENT Colin Hardie	31
DIGITAL DOCUMENTATION MANAGER Lyn Wilson	37
ENGINEER John Andrew	43
FORTH BRIDGES AREA TOURISM STRATEGY MANAGER Karen Stewart	49
GADGET MAN Garry Irvine	55
HONORARY RAIL AMBASSADOR John Yellowlees	61
INDUSTRIAL HERITAGE RESEARCHER Frank Hay	67
JACK-OF-ALL-TRADES Jenni Meldrum	73
KIOSK OWNER Francesco Agus	79
LOVEBIRDS Paul Ward and Meghan Crawford	85
MODEL MAKER Michael Dineen	91

NETWORK RAIL ASSET ENGINEER
Jamie McLaren   97

OUTDOOR SWIMMER
Gina Bees   103

PRINCIPAL TRANSPORT CURATOR
Louise Innes   109

QUEENSFERRY COMMUNITY STALWARTS
Karen MacGregor and Terry Airlie   115

ROYAL NATIONAL LIFEBOAT INSTITUTION CREW MEMBER
Julie Dominguez   121

SKIPPER
Scott Aston   127

TRAIN DRIVER
Gavin Black   133

UNESCO WORLD HERITAGE LEAD
Miles Oglethorpe   139

VOLUNTEER
Len Saunders   145

WRITER
Elspeth Wills   151

X-MAN
Donald Scott   157

YOUR VIEW ORGANISER
Jordyn Armstrong   163

ZOOLOGIST
Cristín Lambert   169

POSTSCRIPT

Author's Note

IT IS THE crack of dawn.

I pull on my hoodie and tiptoe towards the car, careful not to make too much noise. I've got a bit of a drive ahead of me: Inverness to North Queensferry. At least the horizon is beginning to lighten. My destination: the Forth Bridge and the Your View event raising funds for Barnardo's. Intermittently, Radio 4 keeps me company as I devour the miles down the infamous A9. I'm a bit nervous if I'm honest. There are several reasons for this.

As of today, I am the Forth Bridge Writer in Residence. That's wildly exciting, but also a little daunting. There will be many people to meet, and I can only hope that I will be able to deliver something of value to the organisations which have placed their trust in me.

At today's Your View event at the Forth Bridge, I will interview visitors as they arrive and depart and try to get a sense of what this bridge means to them.

'And while you're there, Barbara, you may have a chance to go up the bridge if we can fit you in,' my contact had told me. My stomach churns. I am famous for my pathetic inability to countenance any kind of height at all – even the attic ladder is a step too far. And yet I nodded at this. What's wrong with me? Have I forgotten who I am?

Once I near my destination, I begin to worry about other, insignificant things too – will I find a parking space? No need to fret: a Balfour Beatty employee beneath the bridge waves me enthusiastically towards their small car park. I am almost blinded by the off-the-scale visibility of his attire, only second to the brightness of his smile. 'I'm one of the volunteers,' I whimper through the window, slightly defensively. I don't think he could have cared less, already directing his cheerful waves at the next passing car.

A lean man in a cap and raincoat waits beside my car, clutching a shoulder bag and an iPad – ah, I recognise him from some of the Zoom meetings I attended: Miles, the Forth Bridge World Heritage Management Group Chairman. I feel better already. By the time I get to the portacabins, the Barnardo's special events manager whizzes by. Wiry and fiercely cheerful, she weaves her way through the crowd and introduces me to what feels like the population of a small country. I do what I do best: reach for my notebook and begin: 'Hi!'

If the total strangers before me are unsettled by my slightly deranged smile and erratic gestures, they don't show it. I continue: 'Have you just come off the bridge? Was it amazing?' The answers are always affirmative. I move to the heart of my task: 'What does the Forth Bridge mean to you? Have you got any special connection? Any stories? Anecdotes?'

And so it begins. Shirley from Livingston remembers driving under the bridge on her second date, listening to music and discovering all the things she and her then-boyfriend had in common. 'It's just inspirational, this bridge. I marvel at the engineering, but it's my thinking space too. I sit under the span when I have a problem or there is something on my mind. I've waited years to do this, to go up!'

Dunfermline's Davie O'Donnell works for Network Rail and is often seen at Waverley Station in a top hat. Today, he has returned to the Forth Bridge to lend a hand. However, Davie has his own stories to tell too: He had his distinctive mass of white hair and his beard shaved off for charity atop the Forth Bridge as a fundraiser for Macmillan cancer support, by none other than Rangers legend Mark Hateley – not a day you could easily forget!

The pages in my notebook are filling up fast. According to my elderly interviewee recalling memories of Sunday school trips from Bo'ness to Kinghorn, the highlight was winding down the windows of the train and throwing out pennies for luck over the water. Another lady recalled her four-year-old sister throwing out the entire contents of her fluffy purse – all her holiday money for Scarborough was gone – to her mother's horror!

AUTHOR'S NOTE: BARBARA HENDERSON

Kirstie had won a ballot to go across the Forth Road Bridge in a minibus just the day before. 'Quite unusual, I think, to do both bridges in a weekend. We got a tour and even walked in the cabling! And today I'm going up the Forth Bridge!' Soon after, I meet self-confessed adrenaline junkie May Macleod who abseiled from the bridge as a treat for her 60th birthday, years ago.

A lady volunteer for the North Queensferry Heritage Trust recalls going up the Forth Bridge in the 'old hoist' and regales me with a very entertaining dramatic performance of the experience: 'It was going up in instalments, with a jolt, like this!' she laughs, jiggling. The North Queensferry Heritage Trust is also represented by Garry and Robert Irvine, with their portacabin display of images and documents attracting a steady flow of visitors. I also meet Malcolm, employed by Balfour Beatty in England, who never misses an opportunity to return to the Forth Bridge. He shows me stunning photographs of the bridge's delicate patterns, perfectly reflected in the calm waters of the Forth.

And still they arrive: the engineering enthusiast who travelled all the way from Liverpool, just to ascend the Forth Bridge. A son and his 90-year-old father, taking the hoist together as a birthday treat for the older man. A lady in reflective mood, paying tribute to her late father as she scales the bridge he loved, in memory of him. A young engineer who chose his professional path inspired by the Forth Bridge.

I spend some time interviewing Colin Hardie, the Balfour Beatty Project Lead at the Forth Bridge. His child refers to the structure as 'Daddy's bridge'. 'I was so proud when she said that!' he admits.

Some of the most impressive people I meet are the employees and volunteers: supervisors, site workers, scaffolders (can you imagine building a scaffold on that bridge!), safety inspectors and the Briggers, a local heritage group who ably demonstrate the bridge's impressive history, including a hands-on riveting demonstration.

Finally, the time has come. 'Looks like you are going to get a chance to go up, Barbara, alongside the other volunteers.' I am ushered to the safety briefing, kitted out with a fetching yellow vest and a stylish hard hat. No turning back. I am so grateful for my cheery sidekick Gillian who coordinates the region's fostering for Barnardo's. If she isn't scared, then I need to get a grip of myself. How hard can it be?

The hoist feels substantial enough, and in any case, I am too far gone now. Fear may have a hold on my mind, but I am not ready to embarrass myself in front of my new friends! I am going to stand on top of this bridge if it's the last thing I do...

Granted, my legs wobble a little as I emerge out of the hoist, but the overwhelming sense is one of space and freedom. The scale of the structure is immense, and the thought that Victorians built this without the aid of computer calculations and modern technology leaves me all but speechless. I am astounded that Berwick Law and Edinburgh Castle are clearly visible from here. There is a photographer, precariously balanced on a raised platform, ready to immortalise the moment, and I strike a pose.

It is almost dark by the time I retrace my steps to the car. As I chase the darkening clouds northwards, I ponder my luck. What a privilege, to be among those people, in that place, on this day.

Of course, this was only the beginning. Over the months that followed, I have been lucky enough to discover that the Forth Bridge is so much more than a structure of paint and steel. It is a community, every bit as interconnected and interdependent as its famous red struts and lattices. The beginnings of an idea began to form.

With the support of Luath Press and now teamed up with the astonishing photographer Alan McCredie, I set to work. Oh, there were possibilities aplenty – hundreds of people were suggested as potential subjects. But what I wanted most of all was to showcase the community on and around the bridge in all its colour and variety. That's when I had the slightly madcap idea of an alphabetic structure, a choice I was to regret periodically over the months spent interviewing and writing. However, this self-imposed contrivance forced me to look harder, cast the net wider, think and then think again. Is my selection comprehensive? Absolutely not! Let's face it; it isn't even logical.

However, I am immensely proud to introduce the Forth Bridge

community in 26 chapters. These are people who work on the bridge, promote the bridge, protect the bridge, live by it, encounter it every day, have played a significant part in its story, and much more besides.

In that sense, we are all made from girders, interdependently criss-crossing each other as we remain riveted to the structure which inspires us.

Barbara Henderson
September 2023

Photographer's Note

AS A PHOTOGRAPHY student in Edinburgh I regularly made trips to the Forth Bridge in the hope of taking an iconic photograph of it. This continued for years and I was never happy with my efforts. One afternoon at college in the print finishing room outside the darkroom (yes it was that long ago) I watched as a large print slowly inched out of the processing machine – what I saw appearing before me was the best photograph I had ever seen of the bridge, taken by a fellow student. The feelings of jealousy were quickly replaced by a feeling of immense relief – I was never going to better his image (I never have) so I felt I could now stop my slight obsession with the bridge. I told myself I was done and that I wouldn't be photographing the structure again.

Fast forward quite a few years and Barbara got in touch with me about a book project she was planning on the Forth Bridge. Would I be interested in doing the photography? I remembered my solemn vow to myself to never photograph it again and promptly broke it. Like all my best principles, it lasted only until the exact moment it was put to the test.

The actual process of carrying out the photography for the book was a joy. I spent a lovely summer visiting the Queensferrys North and South (a divide that to some is on a par with North and South Korea) and met and photographed some incredibly interesting people. As Barbara has written, they all have their stories to tell about their relationship with the bridge and each and every one of them was incredibly generous with their time. Of all the genres of photography it is people and portraits that I find the most rewarding and this was a project that I enjoyed from the very first portrait to the very last.

The photography took a few months to complete and was immensely rewarding. I was allowed access to the very top of the bridge (not scary) and also down to below track level (considerably more scary). I was part of a rowing boat crew that took me directly under the bridge from South to North. I sat in the cockpit of a high speed train, and in the wheelhouse of the *Maid of the Forth* boat. I watched as a huge aircraft carrier slid gently below the north span of the bridge. Every visit felt like an adventure.

Early on I decided that I didn't need to have the actual bridge in every photograph. Its presence was always felt though. It is a structure that draws people toward it – things happen around it and people modify their behaviour because of it. It looks different from minute to minute depending on the weather and the light. It can look both immensely solid and achingly delicate depending on your viewpoint. It bridges two coastlines but is also a bridge through time. Whenever I see it I remember my dad telling me how, as a young boy in Rosyth during the war, he witnessed the first German bombing raid on British soil when they attacked the bridge and the port in October 1939. It feels like it has always existed, and always will, and is quite simply one of the most remarkable pieces of engineering anywhere on the planet.

To conclude, I want to offer a huge thanks to everyone I photographed for being so patient and accommodating, to Barbara for allowing me to have such fun, and to the bridge itself for being immense in every way. The pleasure though, was all mine.

Alan McCredie
September 2023

The Forth Bridge is a globally important triumph of engineering, at once structural and aesthetic... It represents the pinnacle of 19th-century bridge construction and is, without doubt, the world's greatest cantilever trussed bridge...

From the successful nomination to register the Forth Bridge as a World Heritage Site

ARTIST
Gordon Muir

MADE FROM GIRDERS: OUR FORTH BRIDGE

'I'M AN ARTIST. With a small 'a' that is,' Gordon Muir begins with a grin. As the designer and sculptor of the memorials to those who died during the Forth Bridge construction, his life and work are closely intertwined with the girders of the iconic bridge. Despite this, he is not easily defined, a wiry mixture of irreverence and intensity. 'If you're an artist, you can apply that to any discipline really,' he finishes.

Having spent the last hour in the presence of the raconteur, I can definitely see what he means. Raised in the Borders town of Hawick, at the time the epicentre of the cashmere industry, he departed for art school in London and soon shifted his focus from Fine Art to Graphic Design. Following his degree, he embarked on a world tour, learning from artists around the globe. 'Wood block and paper making in Japan, lithography in the States – I should have kept travelling really. I returned to Scotland to begin work, but I was far too young. Through a friend, I became involved in the music industry and ended up managing a band from Zimbabwe called the Bhundu Boys.' Under Gordon's direction, the musicians enjoyed considerable touring success, including supporting Madonna at Wembley in 1987. Eventually, Muir moved back to the South Queensferry area, where he has been based for 25 years, to work with friend and landscape architect Paul Hogarth. 'If you live in Queensferry, you live and breathe the bridge, you can't help that. For me, the Forth Bridge is such a huge metaphor for connecting people. It's what happened for me!'

Being steeped in Zimbabwean music and the Bhundu Boys' distinctive Jit style, Gordon's teenage son formed a band with friends: Bwani Junction. 'One of my most memorable Forth Bridge memories is of standing on top of the North Tower at 8am. The clouds were just dissolving beneath us, and my son and his band were playing happy birthday for me. It was the awesomeness of the spectacle, but also a family thing at the same time. At the time there was only a rickety old lift to take you up there, you know!'

Gordon's work as an artist has touched on the Forth Bridge in several ways over the years. 'I collaborated on some of the signage around South Queensferry, for a start. Then I did some work on

the iconic human cantilever picture, creating a sculpture which was going to be turned into a full-size bronze – so far, it hasn't come to anything! I was also one of the first people to don a headset for the Virtual Reality 3D experience – in a small room in Stirling, where I walked around the girders of the Forth Bridge, virtually.' He hesitates, before adding: 'I fell off actually!'

However, surely Gordon Muir's most notable work associated with the bridge are his two memorials, facing each other across the waters of the Forth. He recalls: 'I joined the group which was going to appoint an 'Artist-in-Residence', working towards creating some sort of memorial. After making little progress, we felt it may be best to have something tangible to show such an artist, so I offered, free of charge, to create a design. The idea was to appoint a stonemason to make it, out of a special stone to be imported from India. The budget was not really coming together. However, Network Rail's refurbishment job was reaching its conclusion and they offered us 95 per cent of the money if we could deliver the memorials to tie in with the completion of the job – a ridiculous deadline, basically determined by Alex Salmond's availability to unveil them on the day.'

The team's original plan had to be abandoned: 'There was no way it could be made by a stonemason in the time – the stone had not even been ordered! The only way to do it now was casting it in bronze – and for me to work like a Trojan! I had to do everything: the full-size clay model, the lettering – I was literally sleeping at the foundry. We have pictures of my son and his bandmates working there, helping me in the middle of the night. And we did it! On the morning when Alex Salmond unveiled the memorial, there was a reception at the Orocco Pier in South Queensferry.' He pauses. 'I am proud of that work. Very proud, for my part in that story, highlighting the shocking lack of concern for the lives of ordinary workers in Victorian times.'

Muir's remarkable way with images also led to another opportunity to celebrate the Forth Bridge's heritage – he joined the team of researchers working with writer Elspeth Wills to document the lives of ordinary Forth Bridge workers, an effort which would lead to the book *The Briggers*. 'We approached the National Library of Scotland at the time. They were in the process of digitalising many images, and somehow, someone saw fit to give me access to the Evelyn Carey collection of glass-plated images. The digital scans were huge in format, so I was able to see every plook on people's faces. These Forth Bridge pictures hang in all the pubs around Queensferry, but until you zoom in you don't see that there are men in them, all suddenly visible. Once you look closely, it's like a scene from *Paint Your Wagon* – the ridiculous clothes they used to wear, you know, considering where they were working, with their tweed suits and their deerstalker hats and their handlebar moustaches. And I realised: *I am looking at men and boys who have never been seen by their own family. I'm the first one to see these men in over a hundred years.*' He laughs apologetically. 'Even as I'm talking about that, the hairs on my neck are standing up. It was astounding!'

On a lighter note, Gordon created his own version of the Forth Bridge after his son's band's request to film a pop video on the actual Forth Bridge was declined by officials. 'We filmed the whole thing in an engineering yard, pretending they were on the bridge, with the shots of the girders made out of balsa wood and polystyrene. We even managed to borrow a train and mocked up emergency stop handles to recreate the scene of a train stopping on the bridge. It was pretty convincing, I think.'

BLUE BADGE TOUR GUIDE
Andrew Simpson

'WE STEPPED OUT onto the decking, and all you see is sea and bridge. A LOT of bridge!'

Born in St Andrews, Blue Badge tour guide Andrew Simpson, also known as the Lucky Sporran Tour Guide, never imagined himself living on North Queensferry's water's edge. At the time he and his husband, children's author Justin Davies, viewed the property, both were working as long-haul cabin crew out of Heathrow.

'I suppose I had always hankered back to Scotland a little. When I saw the house for sale during a brief visit to my parents who by then lived in Limekilns, I asked my dad about it. I knew North Queensferry well, but I could not place the image. "Where is that?" I asked. The next time I was up in Scotland with Justin, I stopped and parked up by the house. We walked past. It was just so unusual. Justin said: "We might have to see it."'

Their viewing took place on a clear day in March 2013. 'It was when we went out onto the decking, WOW. The view was just all-encompassing. The offer was in on the Tuesday and immediately accepted. We moved in that June.'

Living beneath the Forth Bridge, what is that like? I genuinely want to know.

Andrew laughs. 'Well, there is still a lot of standing in the garden, just looking out. And the trains, obviously. The noise is not intrusive though. To me, it's actually comforting. If I'm having trouble sleeping, I might hear the last train rattling across and I know what time it is without checking the clock. The same applies in the morning: you hear the sound of the train above – and you know it's 6.40am. Every so often you even see a steam train going across, which is always exciting.'

Andrew was the first of his family in five generations not to become a captain in the Merchant Navy. 'Perhaps just as well. My eyesight just wasn't good enough. But I have sailed from a young age, and you get such a different perspective of the bridge when you sail under it. My dad worked for Forth Ports all his life and my papa was a navigational pilot who worked out of the pilot station in North Queensferry. He actually remembers being hit by coins from above – people used to throw pennies out of the train

windows for luck! Now, I see the ships going back and forth, a constant reminder of my seafaring heritage. But even as a young boy, I had always loved aviation. I might have liked to be a pilot, but again, my eyesight – not to mention the fact that I wasn't great at maths and physics! I organised my own work experience at Edinburgh Airport.'

Despite having a place at teacher training college, the young Andrew chose travel over teaching. 'I loved it, and I did it for 20 years – plus I met Justin. But there came a point when I thought there had to be more to life than flying. It was my mum who suggested trying for a Blue Badge tour guide qualification.' The training for tour guiding is more rigorous than most of us would assume: a two-year course roughly equivalent to first- and second-year university studies. Andrew qualified in 2020, the worst time possible, in the middle of the pandemic. Tourism was dead and Andrew dismissed the tour guiding he had studied so hard for. 'It was pointless, no one was travelling! To make matters worse, both of us had decided to leave British Airways amid all the uncertainty. It made sense to take the money and leave, but that was doubly scary because we lost both incomes at the same time. I took a job with HMRC and pretended to be excited about having an office job for six months – I had never had one of those before. Then I realised – I hated having an office job.'

It was the Forth Bridge which changed Andrew's mind about tour guiding. Someone in North Queensferry put Andrew forward to do a piece to camera about the Forth Bridge for the filming of *Landscape Artist of the Year* on Sky Arts. The segment was filmed in the couple's garden. Andrew recalls: 'When it went out on TV in February 2022, I watched myself stand there in my tweed waistcoat and with my Blue Badge on, and I thought, *that is what I want to do*. I needed to go for it.' Within three weeks, Andrew had updated his website and handed in his notice to the tax office. Soon, he had a full summer schedule of trips and tours. Now he takes any opportunity to spread the Forth Bridge love: 'Whenever I can, I bring people down here. The view from here is so different to the view from South Queensferry. Many tourists are surprised that the bridge is so old, and still operational. They are shocked to hear that it wasn't even supposed to be this bridge, and that the first bridge design was abandoned because of the Tay Bridge disaster. Tour guiding is a great job. You get to tell people about the country you love.' He reflects for a moment. 'Turns out that some of the blind decisions you make can be the right ones.'

In between his many trips and cruises, he returns to the old, listed stone house on the waterfront with its crenulated turrets, octagonal rooms, lighting tower and stone spiral staircase. The bedroom was the superintendent's office for the ferry pier and the downstairs lounge the ferry waiting room. One story goes that the second last legal duel in Scotland was fought in North Queensferry. The unfortunate loser's body was initially taken to the Albert Hotel before this was deemed inappropriate – it was a pub after all. Instead, the body was removed to the house where Andrew now lives. Do these stories affect his relationship with the property at all?

'The house may be old, but it never feels creepy or weird, like some old houses can, whether you believe in ghosts or not. No, it has a unique, content feeling to it, here by the water's edge on the pier beneath the Forth Bridge. My Instagram and Facebook feed are full of images of it. When the sun sets in the west, it turns the Forth Bridge bright red. Early morning sun refractions send shards of light from the structure into the sea which is stunning – especially when there is still a layer of mist on the water. True, sometimes living so close to it, we hear the maintenance works, like sandblasting the access gantry, with its noise and smells.'

He takes a deep breath before finishing, unhurried. 'But that's what you get living next to a structure like the Forth Bridge. It's a living, operational thing.'

CONSTRUCTION SUPERINTENDENT
Colin Hardie

MADE FROM GIRDERS: OUR FORTH BRIDGE

MY FIRST IMPRESSION of Colin Hardie is steel. As senior construction superintendent for Balfour Beatty, he is the project lead for the Forth Bridge structure. Simply put, in terms of bridge maintenance, the buck stops with him. Clad in high-visibility overalls, the 60-year-old removes his hard hat and ushers me towards a cabin beneath the Forth Bridge where it will be quiet enough for an interview despite men coming and going. Shadows of the bridge's many tubes and lattices criss-cross our path.

Back in 2001, the Forth Bridge was due major refurbishments. As a bricklayer, Hardie had experience of temporary structures such as scaffolding, but his previous job was coming to an end. 'The rail bridge needs someone with temporary works experience,' a colleague suggested.

'Nah, that's no' really me – I'm a builder!' was his immediate reply. But he agreed to speak to the project lead and relented: 'Ok I'll come for six months.'

Almost 22 years later, he is still here.

There is real respect in his voice as he recalls his early days as a foreman: 'The first thing that strikes you is the size – even then, I remember thinking *this is enormous*. I was used to working on ice rinks, shopping centres, football stadiums, you know, big things, but THIS was something else. And it wasn't just the bridge itself – I couldn't believe the size of the temporary works – and we were building them from the top down! Up to then I had only built from the ground up. It was a whole new technique, the bridge as a support, using ropes. All of that was new to me.'

'I don't know at what point I suddenly became very comfortable on the bridge. It wasn't days, not a few weeks, not even a few months. Gradually, I became very, very comfortable with it. At that time, during the refurbishments, the place was like a small village, with 400 or so workers for the best part of a decade. We had hoists to the top, and barges below. Doing the work from a boat was new to me too. I had very limited experience of all that. They say every day is a learning day. It really was.'

Colin Hardie eventually became second-in-command for the refurbishment, the one consistent factor throughout the decade-long

process as five project managers came and went. He laughs: 'They all said, "We're not running the show you are!" And that was correct, I suppose. But everybody brought something. It was a team. And as a team, we had a really great bond.'

Hardie recalls only about half a dozen women who were part of the workforce over the years, one or two of the abseilers, or working in the office. 'It was mainly men, I suppose; it was that type of industry. We are still a wee bit behind other industries even now. But you'll find that everyone on the bridge takes a bit of ownership of it.' He tells me of an unexpectedly emotional moment in the early days of his work on the Forth Bridge: 'When I started, I was 37. I'm 60 now, but when I started our kids were young. We used to pop down to Queensferry for a bag of chips, or an ice cream, something like that. Once we said that we were going to Queensferry for an ice cream and my youngest daughter replied excitedly: "Oh, we're going to Dad's bridge!" I remember feeling so proud. I'm not the only one, Michael started working with me at 17. His kids now talk about it as 'his' bridge too.'

Hardie does not expect another major refurbishment on the scale he oversaw in his lifetime. 'We're lucky to have been in the generation we were, and to be part of it. There were loads of social events, even a walking club, and you were there for that period of time. A few people came, saw the work here and thought *naw, we're away*. The majority stayed. Workers here, if they buckle down, they can have 20 years of work ahead of them, and in the construction industry, that is rare. And they learn a lot. We put them through all that training: hoist training, forklifting, even first mate training on the boats, and scaffolding too. Whatever happens next for them, they now have those skills. And because of that longevity, the bridge allows you to develop careers for local young people. You'll get more experience here in five years than in decades of building houses. At the end of the refurbishment period, people jokingly said: "It only took seven years to build it and it's taken you ten years to repair and paint it." That's true. But it's because in Victorian times, they never had all the health and safety rules like we do now. With a bridge like this you don't get any second chances, it's as simple as that.'

Hardie looks out over the water from the small cabin, with distant voices dulled by the waves. He reflects: 'Truth is, I wouldn't have been here for as long as I have if I didn't love it. There's not another bridge in the world that's like this. I'm not soppy about it. But it's a fantastic place to work. You won't get better, you genuinely won't. Don't get me wrong, it's hard. In the summer you get the heat belting off the steel, in the winter you get −14 degrees with the wind chill. It's hard, but it's rewarding – and not just financially rewarding. You're contributing to something.'

DIGITAL DOCUMENTATION MANAGER
Lyn Wilson

IT IS 1AM in the morning and Dr Lyn Wilson begins her night shift on the Forth Bridge, while no trains run. Her team's activity will finish around 6 or 7am, but first there is work to do: operating a bespoke railway bogey scanner on the track to capture 3D spatial data of the bridge, with off-limits access to all parts of the structure. When the sun rises, the views are a once-in-a-lifetime experience. 'I am ok with heights actually, but there are some real heart-in-mouth moments. Some of the areas we accessed are made of a mesh, effectively. From the suspended walkways beneath the track, you look down and just see water.'

What brought her there, you may ask.

As Digital Documentation Manager for Historic Environment Scotland, Lyn was the woman who oversaw the digitalisation of the Forth Bridge – in its entirety.

'It really came about as a result of the UNESCO World Heritage nomination. The Forth Bridge was creeping up the agenda, and we wanted to document the bridge, with a view to help the UNESCO inscription process,' she explains. 'I came from a heritage science background, with a PhD in Archaeological Science. In 2007, Historic Scotland bought a laser scanner, and the process of 3D digital documentation became my thing. It was an area which really exploded, and it has been an exciting and fast-moving journey. There was always new hardware, new software, or both – it keeps you on your toes, that's for sure. The Forth Bridge was a massive project in terms of data management, by far the most challenging we have ever undertaken. Let's just say it took a while. You spend months in the field and then you double that time in terms of processing to make the end product as accurate as you possibly can.'

Lyn Wilson is a passionate and engaging communicator and even as a comparative technophobe, I can't help feeling excited about what she has achieved with her team: a detailed, faithful, as-built digital rendition of the Forth Bridge. I ask her to explain what happens.

She hesitates, perhaps wondering where to begin. 'Basically, you need to capture lots of digital data and photos which are then combined to create a 3D model. It's a fast-paced and ever-evolving

process, but we need a range of technologies to achieve an accurate 3D representation: laser scanners, digital single-lens reflex cameras, 3D surface geometry software, and reality capture software. Our 3D data has XYZ coordinates and when combined with structure from motion photogrammetry, creates a super accurate photorealistic 3D model of the bridge.'

It sounds complicated to me. 'But it's worthwhile,' she argues, 'because of how that data can be used. It's a huge investment in terms of time, effort and budget and it's really important that the data is used well.'

Lyn's journey in heritage digitalisation began with *The Scottish Ten*, an ambitious plan to digitalise World Heritage sites all over the world. It saw Lyn standing on Mount Rushmore, hanging on to the Sydney Opera House's sails and visiting India, China and Japan. 'It really was a life-changing experience, but it also pioneered the power of combining heritage and politics in a really constructive way. You could describe it as soft diplomacy – supported by the Scottish Government, we gifted the data we gathered to our international heritage partners. These collaborations taught us all so much. After all that, we thought: we can apply that knowledge here at home. Let's do the Forth Bridge.'

'We secured funding and began with a pilot phase of the Fife cantilever, working with our partners at the Glasgow School of Art. It was mainly over land, but still very challenging in terms of complexity. It's all very organic – there isn't a straight line on that bridge, honestly!'

By 2015, Lyn and her team had tested and evaluated their methodologies and could begin in earnest. 'Access to all areas was incredibly important,' she emphasises. 'And physically, there was a lot to do. Our scanners can't see through anything, nor behind anything. There is no substitute for capturing data and taking images from all angles, from underneath, from the top. We had rope access teams abseiling, we had boat-mounted scanning systems, we took images from the track itself. The goal was to have no missing areas. It was a long process from start to finish.'

I am curious about the applications of such a project. What

could such an accurate 3D model be used for?

Lyn smiles widely. 'There are so many uses. Working with learning experts, we created the Go Forth learning resources, for example – a game-style set of learning resources which are now available to every school child in Scotland. That in itself was challenging – working with this type of complex 3D data requires terabytes of storage, and schools simply don't have that. How could we make this accessible? We commissioned an education expert to storyboard something that could work in a school. The complex 3D models had to be distilled down and the result is a set of interactive resources we are really proud of, including a section on teaching primary children how to code. We launched them with John Swinney, the Education Secretary at the time, in 2018.'

I am deeply impressed because I am familiar with Go Forth. I used the resources myself during the research phase for my Victorian-set book about a brigger boy, and I tell Lyn so.

'Oh, but there is so much more. It's an accurate record, an as-built snapshot in time. It is vital to have this to be prepared for a worst-case scenario, a disaster of some kind. This data helps us make informed, evidence-based decisions on conservation. But it can be useful in the here and now too: the Network Rail team have access to the 3D data as they devise ways of electrifying the track or install new lighting. The Navy have used it to check clearance heights for large ships, too. But there is also the fun aspect: the visitor experience project in development, where people can enter the 3D Forth Bridge environment wearing a virtual reality headset. Of course, it doesn't compare to actually going on the bridge, but it's an alternative for those who are unable to do so, for whatever reason. And don't get me started on the potential for gaming.'

Now working as Historic Environment Scotland's Head of Programme for Research and Climate Change, she is rightly pleased with what her team has achieved. 'Digital is no substitute for the real thing. But it helps heritage go further and reach wider – and to connect with audiences it wouldn't reach otherwise. And yes, I am so pleased to have played my part in that.'

ENGINEER
John Andrew

'I'M JUST OLD enough to remember the ferries going across the Forth, before the Forth Road Bridge,' John Andrew laughs. 'Our family was always travelling from Edinburgh to Fife for holidays, and our Sunday afternoon jaunts were either to Arthur's Seat or to South Queensferry. "What's that big red thing?" I'd ask as a boy and thought to myself that it would be wonderful to stand on top of the Forth Bridge one day. Little did I know…'

John Andrew's grandfather had worked for Cruikshank's Foundry in Denny, the company that had made castings for the Kincardine Bridge. Perhaps bridges were simply in the blood. But the young John Andrew wasn't one to look back to the past – his aspiration was to press forward. He became an apprentice engineer and following a secondment joined Balfour Beatty, the construction company which now looks after the Forth Bridge.

Rising quickly, he became a project manager working on a range of projects and later was responsible for business development and tender opportunities. 'It's all about relationships, isn't it?' he says. 'One day I got a phone call from Railtrack, the precursor to Network Rail. I had just overseen the restoration and the floodlighting of North Bridge in Edinburgh with its cast-iron façade, including all the networking and relationship building associated with the project. North Bridge was built by Sir William Arrol, the self-made world-leading engineer who had masterminded the construction of the Forth Bridge. Now these Railtrack guys wanted to talk to me about the restoration of the Forth Bridge. I was the entry point for Balfour Beatty's partnership with the Forth Bridge.'

The restoration work was a tendering opportunity. 'Oh, and by the way we're going to the bridge,' the Railtrack caller finished. Not long after, the boy John Andrew had come full circle: he was standing on a 2-ft-wide platform on top of the 'big red thing', albeit in very exposed conditions.

With the contract won for his company, John set to work. 'This bridge starts to get inside you, you know?' he explains as he describes the next chapter of his working life. The engineer had become the Forth Bridge storyteller – fielding media interviews, appearing on television, giving presentations and working on articles to tell the

story of this remarkable structure as it was preserved for future generations. He even secured a guest slot on Blue Peter. During the filming, the Forth Bridge was presented with a Gold Blue Peter Badge, the only inanimate object to have been awarded this significant distinction.

On completion of the decade-long restoration project, John Andrew assisted Historic Environment Scotland with the submission to UNESCO for the historic World Heritage Inscription for the iconic bridge. This came with its own unique opportunities: 'In 2014 I visited Australia and stood on the Sydney Harbour Bridge – that was another bridge off my bucket list! In 2016, I got another call from the railways, by now Network Rail. Would I speak at a conference in Canada on their behalf? I visited the Pont de Québec, which is a cantilever bridge just like the Forth Bridge. It needed restoration. In fact, they were facing exactly the same problems that we had with the Forth Bridge: rust and deterioration. It needed restoration work. I stood on top of that one too,' he laughs.

With his time at Balfour Beatty at an end, John Andrew was elected a Fellow of the Institution of Civil Engineers. 'I kept up my involvement with the Forth Bridge. I then became a project manager with Transport Scotland and got involved with the Forth Bridges Forum. The truth is, every time I go in one direction, I get pulled back. The Forth Bridge has been the one constant during many stages in my career.'

It is perhaps no surprise that he became a key player in the arrangements for the 130th Anniversary Forum for the Forth Bridge in 2020 and increasingly became passionate about maintaining good records and information. Researching the background of his beloved bridge and meeting what he calls affectionately 'mad collectors' became a new focus. 'It was about spreading the word. I was never a historian, not into history at all. But now I can't get enough of it. It's not about me, it never was. It's about giving something back, giving the bridge its place, you know?'

He now helps as a curator at the ICE Scotland Museum at Heriot-Watt University. 'We have an original riveting machine, invented by William Arrol to assist with the output in the placing of some of the 6.5 million rivets used to keep the Forth Bridge together and to address a specific problem. He was a self-made man, and that is inspiring. I'd like youngsters today to realise that it's not all about university. Learn a trade, hone a skill, become an apprentice. The Forth Bridge is an icon of Civil Engineering, built by a guy without a degree.'

So how does he reflect on a career so closely intertwined with Scotland's greatest man-made wonder? John Andrew is thoughtful for a moment.

'The first time I stood on that bridge was 2001, just to be involved, to feel part of a team. You develop relationships, even friendships, when you work together like that, and when you face challenges together. We love this bridge. The truth is, I never get that anxious about things. Engineers are problem solvers. If you panic, you don't get things done.'

It is evident that John Andrew is not remotely interested in slowing down. In addition to his continuing involvement in Forth Bridge and other transport-related matters, he puts his considerable problem-solving experience to good use elsewhere. He can be found drawing, designing and building sets for live theatre at the King's in Edinburgh, or in charge of logistics and production at St Giles' Cathedral where he oversaw the arrangements for the Service of Thanksgiving and Vigil for the late Queen Elizabeth II in September 2022.

'It's the same skill set: problem-solving, planning, resources, logistics. I know what engineering has given me, and I am so grateful.'

FORTH BRIDGES AREA TOURISM STRATEGY MANAGER
Karen Stewart

'I am terrified of heights I really am!'

Karen Stewart was relatively new in post when, as the Forth Bridge Area Tourism Strategy Manager, she was invited to lend a hand at the annual Barnardo's fundraiser. 'Lots of people pay to go up to the top of the bridge. I was writing their names on the certificates; that was my task on the day. Suddenly one of the team came up to me and said that the volunteers would go up the bridge together. "Ha! No thanks, absolutely not!" I said, but despite my protests, I found myself coaxed into the end of a safety briefing and ushered towards a hoist. My overriding memory is the ill-fitting hard hat. With every picture that emerged afterwards, I think *I wish I'd got that on straight!*'

The Forth Bridges Area Tourism Strategy is all about realising the tourism potential of three major bridges that span three centuries of engineering innovation. Unlike many others listed in this book, Karen is a relatively recent arrival to the Forth Bridge community. 'I was in charge of marketing at a small Fife distillery before but was made redundant during Covid. Now, it was time for a change! I saw the advert for this post and applied. It was the first interview I attended in my slippers – the world had gone virtual by then. My top half was presentable, but no one could see my feet,' she laughs.

Having commuted from West Lothian to Fife for her previous position, Karen was very familiar with the sight of the bridges. 'I watched the Queensferry Crossing being built – and now I have different feelings about them all. The Forth Bridge is this amazing constant, and I love the heritage and the history of it all. But the Queensferry Crossing is this sleek, modern and elegant creature. The Forth Road Bridge is perhaps most relevant to me in my job: as the only bridge that people can actually walk and cycle along, it forms part of the Forth Bridges Trail, and walkers get such a wonderful view of the other two bridges from it.'

Karen's work is largely project based. She explains: 'Jobs like mine tend to come with shorter term contracts, so you have to aim to make a difference within the timeframe you know you are going to be there. That is one of the downsides, I guess – some things just take longer to put in place, and you can't really tackle

those. I have to focus on things I can actually deliver: signage, the Forth Bridges Trail, the website, and some strategies to encourage sustainability and to bring the area together. It feels good to do something positive in the wake of the pandemic. Saying that, a lot of my time is spent on social media, both looking for things I want to say and responding to others. There is so much photography content out there, and it keeps me on my toes finding more unusual angles. I aim for a mix of business news, images and interactivity. The top performers in terms of content engagement are definitely aerial photos – you know, when people fly into Edinburgh Airport and snap the three bridges from the window? Those pictures do really well!'

Karen also attends a lot of meetings as part of her remit. 'There are a lot of different groups and organisations involved with this project, so I attend a lot of meetings to share information and gather feedback. I'm so impressed with all the voluntary organisations and the people who are so generous with their commitment and time. It feels as though I am working for a community, not an organisation.'

Karen Stewart's voice is animated when I ask her about some of the high points of her tenure so far: 'It's great when something you've been planning comes together, like the Forth Bridges Trail where interested visitors can follow a circular route with wonderful views. I loved when the new website was ready, and when a recent film about the area aired, which I collaborated on. I was buzzing – it was just so well done, and they had got the tempo of the place right – it felt slow and considered, in keeping with how things feel here. On the other hand, there can be frustrations when I'm battling to make things happen. The most annoying obstacles can be random things. For example, you can't just hammer a sign into the ground – it has to be approved and there is paperwork. Half the time it's trying to figure out what you actually need to know. But I'm an optimist, so I tend to take a leap of faith and stress about the detail later.'

So, if she had one wish, what would she ask for?

She thinks for a moment.

FORTH BRIDGES AREA TOURISM STRATEGY MANAGER: KAREN STEWART

'My biggest wish is for a physical place where our visitors can find out more about the amazing heritage of the Forth Bridge and its communities. The proposed Forth Bridge Experience will cater to thrill seekers who want to don a jumpsuit and a harness and walk up – but I'd love to see a visitor centre to appeal to a wider audience, too, a venue which could bring the many stories of this place to life. That would be Utopia!'

Karen's enthusiasm is evident when promoting the iconic structure. 'It just feels great to tap into something people already love, and with such longevity. The Forth Bridge is for every generation, locals and visitors alike. The UNESCO World Heritage status is such a source of local pride. And isn't the Forth Bridge beautiful? Whenever I look at it from a new angle or in different weather, there is another wow moment. But it's built for a purpose, and it serves that purpose to this day. I think the designers created something instagrammable before anyone knew what that was!'

She pauses. 'You know what? It feels weird talking about myself like this. I am supposed to be anonymous in my role – the bridges are supposed to be the story, not me!'

GADGET MAN
Garry Irvine

THE NICKNAME SEEMS apt – throughout his life, Garry Irvine has embraced new technology ahead of most of his peers. He likes to *be prepared* as the Scout motto declares. Not only does he carry a pocket digital camera as well as his smartphone wherever he goes, but he also wears a belt with his pocket camera, spare battery and memory cards, pocket multi-tool kit with pliers and screwdrivers, as well as a torch and a small tin of odds and sorts such as screws, pins etc. 'My father was in the Scouts, and I guess I took the Scout motto to heart.'

Originally from Galloway, Garry remembers his first glimpse of the Forth Bridge on a Boys' Brigade bus outing to Edinburgh in the mid-sixties. '*What a humungous effort it must have been to build this*, I remember thinking. Even now. what impresses me so much about the Forth Bridge is that there were no computers involved in the construction. All the modern tools we would consider essential – they had none of that. It was a triumph of the intellect of bridge architects and builders. It's truly astounding.'

Much later and by then based in Fife with his family, Garry and his wife Karin joined a mystery car tour as part of a village gala day. He recalls travelling the winding road up Ferryhills to North Queensferry and they agreed: *This is a nice, charming place to live.* Photography had long been a passion, and the village combined a stunning setting with proximity to his work at South Queensferry and with easy access to the motorway and Edinburgh Airport. 'Karin was from Norway, so we had wanted to give our children a home in the countryside. To me, this was the next best thing to Galloway – Galloway was always going to be "doon hame" and beautiful, but North Queensferry had everything else!' he quips. He has been here for almost 40 years.

When his wife died almost ten years ago, Garry began to volunteer for the North Queensferry Heritage Trust where he now serves as secretary, writing the Trust's newsletters and helping with a range of duties. 'I suppose I was a bit lost, and on autopilot,' he reflects. 'I needed to fill my time with something constructive, so I also began to volunteer for the Scottish Norwegian Society in Glasgow, and locally with the North Queensferry Heritage Trust. The Trust

manages the Old Railway Station, which was refurbished back in 2015, and it now houses a gallery, art studio and accessible toilets.'

Best of all, the gallery contains a range of displays and artefacts which tell the story of the bridge and the local area. Over the years, Garry has become a knowledgeable and enthusiastic local guide, despite an unpromising early start: 'I had a terrible history teacher, so I wasn't into history at all,' he concedes before throwing me a snippet of historical bridge trivia, as he often does: 'Did you know that on the opening of the Forth Bridge on the 4 March 1890, the Prince of Wales questioned why there was no permanent station at the north end of the bridge at North Queensferry? A mere 90 days after that, the station was built to serve the local community.'

No, I wasn't aware of that particular fact, but I am determined to remember it.

Garry relishes sharing his love of the bridge with visitors and locals alike, and he volunteers at the world's smallest working light tower on North Queensferry town pier. 'It is 23-feet-tall, so you tower over the old pier, but you are still in the shadow of the 361-foot-high Forth Bridge. It's quite something. The visitors who want to light the lamp have many reasons for doing so, some in memory of others, or for many a special birthday gift. The lighthouse was built in 1812 to help ferries and ships navigate the treacherous tides. This lighthouse's light lasted only five years before Robert Stevenson was commissioned to improve the lighting across the Queensferry Passage, resulting, in 1817, in the present light tower with a fixed light facing its twin in South Queensferry between the harbour and the lifeboat station. I love talking to children about it too – they always go to the heart of things, to the *why this* and *why that*. On one occasion, a child of about nine or ten arrived with his parents. He asked lots of questions; it was exciting for him. But when he looked up at the bridge at the end of the tour, he had "just" one more question as he looked up and pointed to a little house on the bridge – the bothy for the workers. "Why is that bit made from wood?" he asked. It was a good question, as all the rest of the structure is made from metal, from steel! I couldn't answer that but suggested that it was probably easier and cosier to make it from wood. I loved his passion, and it is these typical conversations with our younger visitors that make my volunteering so memorable and worthwhile.'

Volunteering is a considerable commitment – Garry will typically spend a minimum of four to six hours a week on heritage work, but there are compensations: the people he meets and the stories he hears. At a fundraising event, he met an old lady. 'She had lived right under the bridge, or maybe one of her relatives had. Anyway, she told me that they were out in the garden and found a man's head.'

I am speechless (and suddenly a little nauseous) until he elaborates: 'You see, trains were passing over the house all the time, and in those days, you could still open the train windows. This man had literally lost his head looking out the window. True story.'

I had asked for Forth Bridge anecdotes, and I certainly got one from him there. But above all, what shines through is Garry's passion for engineering detail, and for educating young people about their past. He is a local activist, but far from parochial: 'The bridge has two ends, I always say. The sun always shines on North Queensferry, of course, but the whole area is a community around the Forth Bridge. Now we have the Forth Bridges Trail and I help other groups and committees with a common purpose of promoting the bridges and the whole wider area. When designing the Forth Bridges Trail brochure we agreed that as the bridge had two ends, we would cover both "north" and "south" irrespective of the local political set-up. We believe a unified approach works best.' He nods to himself before summing up. 'And I'm keen to help.'

HONORARY RAIL AMBASSADOR
John Yellowlees

'DID YOU KNOW that every railway bridge in the country is numbered? Every single one. With one exception. You've guessed it, the Forth Bridge. Because this one is simply known as THE Bridge.'

'Oh, and did you know Joe Biden, the American President wanted to see the Forth Bridge when he came to COP26? The presidential motorcade was diverted at his request.'

'And, did you know that the Forth Bridge was dedicated as a *Sri Chinmoy* Peace-Bridge by India's High Commissioner to the UK in 1996, implanting into it "the soul of peace"? I was there.'

I would describe John Yellowlees as an encyclopaedic waterfall of Forth Bridge Stories – it's not a trickle; it's a flood! The Honorary Rail Ambassador at ScotRail barely takes a breath, and my hurried scribbles in my notebook become ever more illegible as the pages amass. Born in Edinburgh in 1951, he remembers the diesel trains crossing the Forth, and recalls travelling on a ferry where there was a small knot in the wood, meaning that the ocean was visible below. His father kept an allotment and often sat his young son John down near a train track. 'He left me there to take down the train numbers, as an activity, I suppose. I was smitten with railway travel from a very young age. My mother was friendly with the wife of the local ASLEF secretary, the railway union. I suppose I was very pro-railway, and a bit of an activist, writing letters to the managers even when I was a boy.'

After school, John studied Geography at Cambridge and Urban Design at Edinburgh. The natural route was to become a town planner, but he found himself on a civil service fast-track programme instead, despite his realisation that, on balance, he would rather be in Scotland than in London. His professional life oscillated between the two. Government business was his work, but railways remained his passion until a fateful journey on a sleeper train which was delayed. 'I rang a contact at ScotRail to complain. To my surprise he invited me for a curry, and that's how it started: I began to attend railway events – I suppose I was a useful contact for them too, because I was mixing in this milieu of ministers down south. One day, at five past twelve on a really busy day in my office in London, the phone rang.

I told a colleague to answer it: "Tell them I'm busy." It was John Boyne of Railtrack. They were looking to appoint a stakeholder, but they had too tight a deadline to advertise it externally. They needed an outsider. I remember answering "If you are that desperate, I'll do it." And that was it – I worked on ScotRail's public affairs team from then on. In terms of returning to Scotland, it was good timing; my mother's health was declining and there was no question of a compassionate transfer from my existing job.'

Throughout our chat, John inadvertently name-drops; I think he cannot help it. 'The writer Iain Banks told me that he and his brother used to walk on the bridge on New Year's Day when there were no trains – a bit of a risky thing if you ask me, but perhaps it was a North Queensferry rite of passage, that.' On another occasion, he tells me, the team chose to invite the widow of Kenneth More, the actor who had starred in the film of John Buchan's *39 Steps* (in which the hero – played by More – escapes from a train on the Forth Bridge) to unveil a plaque. Then he recalls the visit of the Mitsubishi chief executive Imada, grandson of the Forth Bridge engineer Watanabe who features in the famous human cantilever image. 'I heard that when the train came around the corner, he cried when he saw his grandfather's bridge. And for the centenary of the Forth Bridge, my wife and I travelled on the train with Prince Edward who was there to turn the floodlights back on.'

However, the one public persona who unifies John Yellowlees' 'two lives' – his civil service and railway careers – is politician-turned-TV-presenter Michael Portillo. 'Back in London, I sat on a Docklands committee which he chaired, and we met two or three times a week for months. I never got to know any other politician the way I got to know Michael Portillo. When he became involved in Great British Railway Journeys, I was asked to come on board as a researcher and advisor. It basically meant standing out of shot and making sure that no equipment got stolen,' he laughs. 'It also meant liaising between the film crew and the train crew and making sure everyone was happy.'

I'm curious: Does he have a top tip for the best way to enjoy the bridge?

He advises me to take the opportunity to take a boat trip to Inchcolm Island on the *Maid of the Forth*. 'Once you're underneath the Forth Bridge, all the straight lines turn into curves. It really needs to be seen to be believed.'

Now he describes himself self-deprecatingly as 'part of the railway chattering classes', but the truth is that the transition from employment to retirement was barely noticeable for John Yellowlees. 'Yes, I retired – but then I immediately took on the voluntary role of Honorary Rail Ambassador, which came with expenses. I was continuing with the same organisation and with many of the same colleagues. Instead of a retirement do we had a relaunch party, with donations to charity instead of a gift for me. I haven't stopped since.'

He continues a little pensively: 'But make no mistake, the Forth Bridge touches something in the national psyche. Yes, we celebrate it, but there is a daunting side to it, too. It can get to you.'

He pauses, before finishing:

'If I had to sum it up, I'd say this: the bridge is bigger than us all.'

INDUSTRIAL HERITAGE RESEARCHER
Frank Hay

RESEARCHER, RETIRED ENGINEER and campaigner Frank Hay remembers it as the end of a nightmare scenario: 'First Minister Alex Salmond was coming to unveil the memorials to the fallen briggers, to mark the end of the Forth Bridge refurbishment, and our memorials had to be ready. Initially, the memorials had been planned as stone monuments, which would have been easier to maintain. But the timescale became a problem, with the delivery of the stone from India delayed. It nearly blew the whole plan out of the water. We urgently needed a plan B!'

'In the end, the sculptor Gordon Muir worked around the clock with Powderhall Bronze to get the memorials finished. Delayed funding had fallen into place at the last minute and the maintenance company Balfour Beatty offered to secure the monuments into place. What an immense relief it was!' Frank Hay's smile is a little strained reliving the memory, but he finishes with his trademark mix of wisdom and humour: 'It's often a bit like Christopher Columbus, isn't it? You head off, not knowing where you are going. You arrive not knowing where you are, and you return not knowing where you've been!'

Contemporary records compiled at the time of building the Forth Bridge cited only 57 deaths. When the local history group, which Frank Hay had joined on moving to South Queensferry, was tasked with finding the names of the victims for the potential memorials, little did he know where it would all lead. 'But I was all in,' he explains. 'I love research. I love engineering, and I love history.'

Even as a little boy, Frank was interested in how things worked. 'I took my mother's vacuum cleaner to bits and oiled it before putting it back together again and it worked better,' he remembers. 'It soon became clear that I'd have to be an engineer. At primary school, my writing was often covered in red lines – bottom of the class in spelling. But as I moved to secondary school, I could be top at practical work – my woodwork teacher in particular was really inspiring.'

Frank's path was now mapped out, and he pursued a career in electronic engineering. It was an exciting time of new developments

and devices, but he left university 'with a head full of theory, and little idea of how to design or test anything.' However, in his first employment, he found himself testing the world's first brain scanners. 'It was fantastic – I worked with an engineer who had been a ship's cook. He was what I would call an intuitive engineer!' he enthuses. 'Soon I moved on to designing aircraft computers. It was the ideal job for me.'

As an engineering enthusiast interested in history, Frank was naturally drawn to the Forth Bridge. 'As history buffs, we tend to look at people like ourselves in the past, don't we? I wanted to look at people who designed and fixed things. When the names for the memorial were sought, I volunteered alongside three others. We looked through newspapers covering the seven years of the Forth Bridge construction. I was fascinated by how people made, designed and tested things then. Wilhelm Westhofen, a Forth Bridge engineer, kept an official record. It's not an easy read, but we found out so much more than simply the names for the memorials. I discovered that I loved, absolutely LOVED researching – especially finding out things that no one else knows!'

It was meticulous work. Frank roped in help from his sister-in-law, Sheila, who was a trained genealogist, and a giant spreadsheet was begun: deaths, newspaper clippings, corroborations – was this person killed on the bridge? Was their death a direct result of the building work? 'It was painstaking – in fact, the process to identify the names of the victims took us longer than it took to build the bridge in the first place!' Frank quips. 'But we unearthed some wonderful research – for example, my son Adam found the records of Dr Hunter, the doctor on site, who wrote his MD thesis on the effect of compressed air on the human body.'

When the team had found the 73 names, another question beckoned: What were they going to do with all their detailed research? 'We decided we did not want to waste it and aimed to write a book. But after months of trying, we had an introduction only – and even that wasn't very good! We enlisted writer Elspeth Wills and designer Gordon Muir to help us do it properly.' The result is *The Briggers*, a research-dense and yet immensely readable account of the ordinary people who created the extraordinary structure. 'I love the early photographs,' Frank enthuses happily. 'What an incredible process it was at the time, with 15 by 12-inch glass plates in a huge camera. How things have changed, eh?'

A keen collector, Frank's house is crammed with Forth Bridge-related memorabilia of the engineering kind. He holds up a battered metal cone. 'Any idea what this is?' he tests me. Of course, I haven't got a clue.

'It's a rivet bucket! I got it on eBay!'

I am baffled. 'How was it listed?' I ask.

'Lampshade,' he replies cheerily. 'I always try to imagine what a non-engineer would think it would be, and then search for that term. I have loads of interesting stuff now: rivets, early postcards, albumen prints, early books – I could furnish a museum with all of this! If only we could get funding to do something like the Tower Bridge Museum in London, or the Titanic Experience in Belfast, that's my dream!'

For now, Frank regularly teams up with Len Saunders to give public talks on the story of the Forth Bridge. 'People are really interested in this stuff, and often really appreciate the talks – and that is so satisfying. The thing is, there is so much more beyond Westhofen. Unseen documents and images are still emerging, and believe it or not, if you look hard enough, you can find Forth Bridge links with Himalayan wooden bridges and even mermaids. Our most recent presentation was to a group of Blue Badge tour guides. But we have visited societies, community groups, church groups, schools and museums – you name a group and we have probably put them to sleep!' he laughs.

Considering Frank's infectious enthusiasm, I must admit, I find that hard to believe.

JACK-OF-ALL-TRADES
Jenni Meldrum

JENNI MELDRUM IS about two years old in an old photograph, taken in front of the Forth Bridge. 'My godfather and his family had fallen in love with Scotland. I am sitting on the roof of his car, and incredibly, I think I remember the moment. I remember being unhappy for some reason – probably with my position on the roof!'

Jenni's long association with the Forth Bridge speaks to the longevity of the structure. Brought up a few miles along the road, Jenni Meldrum has been a South Queensferry resident for 40 years. 'The bridge changes all the time,' she enthuses. 'One minute it's majestically emerging from the mist, the next it glows bright in the sun – it is ever-changing.'

I had sought Jenni's advice for a previous book project, but I struggle to categorise her – her love of the Forth Bridge appears to take many mantles: researcher, collector, antique dealer, social historian, volunteer. She is a jack-of-all-trades, I decide.

Most locals will know her as the proprietor of the Seakist, an antiques shop on South Queensferry's picturesque High Street – so picturesque in fact that it has repeatedly featured in TV programmes such as Antiques Road Trip. Jenny laughs: 'I have been on that programme about ten times – but I think it's all to do with the position of the shop overlooking the bridge.' The format of the show is simple: two antiques experts drive around the country in a classic car, stopping at various antique shops on the way. They buy items which are later sold at auction in the hope of making a profit. Occasionally, the programme features celebrities too. 'The first time they came into the shop, I wasn't really very accommodating,' Jenni recalls. 'They tend to try to beat down the price of whatever object they are interested in. But then I realised that I could never have afforded that type of advertising on the BBC; it was a gift – the camera even panned along the sign of my shop. On top of that, any profit they make at auction goes to the charity Children in Need, so what's not to love? An Eastenders actor came in with them once – and the experts are every bit as nice as they appear on the telly.'

When Jenni started the shop, she wondered what type of business might work here. What might people be interested in? She herself

was drawn to history, and as antiques get people talking about history, she liked the idea of an antique shop. 'But it was not a very commercial proposition,' she quips. 'It was never about making lots of money, and maybe just as well.'

Her personal interest in the iconic Forth Bridge goes hand-in-hand with her work at the Seakist. 'In antiques circles, anything related to the Forth Bridge seems to have my name on it,' she laughs. 'People tell me about it or make me aware, which can be both an advantage and a disadvantage.' So, what type of objects and memorabilia does she stock? I ask. 'I get a lot of postcards and picture plates from various decades. But I particularly love the items which lend themselves to stories. For example, I have a couple of silver brooches with the Forth Bridge on them. Silver is hallmarked, so you can date it pretty precisely. These are from 1890, the year the bridge was opened. I obviously can't prove it, but romantically, I like to think they were perhaps given by a brigger employed on the site to his sweetheart on completion of the work. The jewellery has a feminine angle on the bridge, and I love that it's different.'

As a researcher, Jenni joined local historians in their quest to find the names of the fallen briggers for the memorials. However, she found herself fascinated by different aspects of the bridge than her all-male cast of colleagues. 'The bridge is such a male-dominated subject matter, with its engineering history and all the technology. It was at times tricky to be a woman in that group context, but I think I did add a different perspective to the book we collaborated on, *The Briggers*. First and foremost, I am a social historian, and for me people were at the heart of the stories I wanted to unearth. It all started when I joined the local history group as a new resident here, to get involved with local people. Later, a smaller subgroup was set up to research the names of the victims during the Forth Bridge construction, and I volunteered to be part of that too, alongside three men: Jim, Len and Frank. However, it transpired that four people can't write a book – our perspectives were so different. Elspeth as a professional writer ended up overseeing the process and we became close friends over the years. The research never really stops. I looked at incident books of injuries and admissions in the Edinburgh Infirmary, as well as incidents of crime and drinking surrounding the bridge. All of these offer such interesting snippets of what it was like living here at the time. But you do have to be careful about what you share – many of the people we read about still have descendants in the area.'

Her life, both personal and professional, is deeply intertwined with the girders of the Forth Bridge. What does the structure mean to her?

She hesitates. 'The Forth Bridge has always been prominent in my life, but if I had to sum it up, I'd say it means *home*. If I see it from a distance, I know I haven't got far to go. Throughout my whole life I have viewed it so often, including with friends and family members who are no longer around. But I don't feel sadness when I look at it as it changes and changes again. The Forth Bridge remains. It's a constant in my life and I love it for that.'

KIOSK OWNER
Francesco Agus

FRANCESCO AGUS LIVES in Fife and encounters the Forth Bridge's iconic vista every single day as he crosses the water on his way to work. The lively and cheerful Italian explains: 'Initially, back in 2016, my wife Franca and I opened a coffee shop on the High Street in South Queensferry, a business opportunity like no other. The attraction was the Forth Bridge, of course – it had national and international appeal to visitors, and the coffee shop was right beneath it.'

Francesco's background lies in hospitality. In 1998, his job took him to London where he met his wife who worked in the same sector. 'It is bizarre – both of us are from the Italian island of Sardinia, but we never met each other until we came to the UK,' he laughs. The couple were transferred to the Middle East for a few years with stints in Dubai, Bahrain and Jordan before deciding to return to the UK. 'We wanted to be nearer family,' he explains. 'But London was not for us anymore – too busy, too expensive. It's fine when you are young, and it offers a lot, but things change when you are a bit older.' They joined a tour of Scotland in early 2016 and simply fell in love with the country. A visit to the Forth Bridge was among the most memorable stops. 'That tour was three weeks long and we just loved it. We thought "Why not Scotland? Let's go, let's see!" We were keen to set up our own business in hospitality and viewed a lot of properties in Edinburgh. But when a South Queensferry option came up, we decided to take a look at that too. After all, we had seen the bridge on our tour. When we arrived to view the property, it was a glorious day, an iconic view, a beautiful village. We were sitting right beneath the Forth Bridge and decided to go for it.'

The purchase proved a success – from 2016 to 2020, the coffee shop thrived, and crucial partnerships were growing. Francesco explains: 'We gradually branched out into outside catering – lots of companies and charities were running events by the Forth Bridge, and plenty people worked nearby too. Well, we started delivering our famous lunch bags to events and businesses by the bridge.'

The couple were establishing themselves in the community, but a crucial step in their journey came in 2018. 'Do you remember

the Beast from the East?' Francesco asks me. 'We must have been the only business owners silly enough to drive through the snow to Queensferry and open up. Everyone else was closed. There was a lot of press activity about the storm, and Transport Scotland's response was filmed at the official viewpoint for the bridges. The Minister for Transport was giving interviews and there were reporters everywhere, but their event caterer had let them down. Because we were physically there and our coffee shop was open, they approached us at short notice, and we delivered food and drink three times a day during the crisis to all the emergency workers who were overseeing the essential work on our roads. There was already a kiosk on site, and when the lease came up, because of our strong relationship with Transport Scotland, it was offered to us first.'

Francesco and Franca took over, running the kiosk as an extension of the coffee shop. Ever nimble and inventive, they continued their strong partnership with the Forth Bridge when they designed a chocolate rivet to sell at the 2019 Your View fundraising event, based on an original Forth Bridge rivet. 'I was just brainstorming with Michael Dineen, one of the Transport Scotland guys at the time, and we wondered if we could do something special for the event – and a large chocolate rivet seemed perfect. The metal rivets which were used at the time are actually very heavy, and the men had to hammer them into place when they were red-hot – I have a lot of respect for the workers who did all that, 150 years ago and high above the water.'

When the Covid pandemic closed the coffee shop, the couple concentrated their efforts on the kiosk. Its location is ideal, beside the official viewpoint for the bridges and with plenty of outside seating. Tour buses stop here all summer season, between April and October, while cruise liners dock by the nearby Hawes Pier. I ask Francesco about a typical day.

'We normally start early, between 6.30 and 7am, prepping all our products,' he tells me. 'Everything is fresh, and our menu includes Italian options as well as locally sourced, fresh Scottish ingredients. We even have our own coffee blend. By 8.30am we are ready to rock 'n' roll!'

The kiosk's regular customers are first: employees of BEAR Scotland, Transport Scotland's control room and the education centre pop in to collect their morning coffees and breakfasts. The first lot of buses appear soon after, and tourists arrive throughout the day. 'Honestly, I can't tell you how many nationalities we see. But they all do the same thing first: they go and take a picture of the Forth Bridge. Every single one of them does that!' The last of the tour buses may pull in for a pit stop around 4.30pm, after a hard day's sightseeing.

It is a joy to speak with Francesco, mainly because he is a bit of a storyteller. For example, he recalls a charity abseil from the Forth Bridge when several participants froze in fear halfway down the bridge. 'People got stuck between the top and the bottom,' he tells me with glee. 'We could hear the instructors shouting: "Don't worry, you're safe!" Over and over. It was a big fuss over nothing!'

As someone with a healthy respect for heights, I am not sure I agree with his assessment. However, there is no doubt that its location places the couple's business at the heart of what is happening around the Forth Bridge. Most recently, the kiosk took centre stage when the UK's first autonomous bus service was launched. The 14-mile route is thought to be the first in the world to use full-size autonomous buses which effectively drive themselves, including crossing the water here, with unparalleled views of the Forth Bridge. All the marketing, including a day of live coverage across national and international media, took place at the viewpoint beside the kiosk. Francesco remembers: 'Scotland's Transport Minister, Kevin Stewart, was there to give interviews on radio and TV, including the BBC. It was incredible press exposure that whole day: our viewpoint and our kiosk were live everywhere.'

I wonder what is next for Francesco, Franca and their business. 'All I know is, we recognised a great business opportunity, and it has given us a lot back. Most of the opportunities that have come our way though, let's be honest, have come because of the bridge. And for that I am grateful.'

LOVEBIRDS
Paul Ward and Meghan Crawford

'WE HAD HAD our first date beneath the Forth Bridge,' begins Paul Ward. 'That was during the pandemic, and we met at the Hawes Inn before going for a walk around South Queensferry. Even for the next few dates, we seemed to always be able to see the bridges from the places where we were. And then Meghan surprised me with tickets to go up to the top of the Forth Bridge – and about a month later it occurred to me that this was going to be the perfect moment to ask a certain question.'

I had been looking forward to this interview immensely – who does not love a good romance? Paul and Meghan, a particularly cheery pair, got engaged on a windy September day on top of the North Tower of the Forth Bridge. Paul recalls: 'I began googling. *What does an engagement ring look like? How do you buy one? What should it be made of?* I also spoke to Meghan's parents and even enlisted the help of my running partner for advice.'

Secrecy was paramount. In an ever more clandestine effort, Paul contacted Meghan's work and secured a Monday off for her so that the couple would have a long weekend at their disposal. He covertly booked a hotel in South Queensferry. But despite his best efforts, the process was far from straightforward. 'Well, it was Meghan herself who threw a spanner in the works – she sent me off to collect a light fixture she had bought online, so I had to suddenly fit my plans around that. Anyway, I spun a yarn. As a keen runner, I told her: "I want to go and watch a run – you wouldn't enjoy it." Instead, I dashed off to drop overnight clothes at the hotel, book a restaurant and – crucially – make sure that there was champagne!' I tease that he must have been very confident of a positive outcome. He laughs at that. 'Either that, or I'd have a nice quiet night, and a bottle to myself!'

It was time. On the big day, the ring box was safely stored in one of the many pockets of Paul's enormous winter jacket. 'But then I read that we weren't going to be allowed to have anything loose on top of the bridge – everything had to be on a lanyard, or strapped to your body, and I thought *erm, I do have something quite important here that I'd like to pull out on top of the bridge.* I decided that I should probably speak to the staff and explain, but

they were relaxed: "Lovely! Yes, on you go," they said.'

All this time, Meghan was none the wiser. 'I had no inkling, absolutely not,' she insists. 'But the weather had turned. We were in the safety briefing, and the Network Rail people were telling us that they were going to do their best to get us up the bridge. But there was a chance it might not happen; the winds were so strong.' Meghan noticed that her boyfriend looked tense in the hoist and began to feel bad for booking the trip. She stroked his arm: "I thought you were fine with heights. Are you all right? You look really worried."'

Of course he did, haunted by nightmare visions of a treasured ring bouncing right through the cracks of the platform and plunging into the depths of the Forth below.

'As soon as we were out of the lift, we were blown away by the view and what you could see. I turned to a fellow passenger, handed him the lanyard with my phone and asked if he could take a picture of us. Out of the corner of my eye, I saw Paul frantically patting himself down. When I turned, he was on one knee!' She raises her eyebrows.

Paul supplies the context: 'Yes, when Meghan handed her phone over, I knew that it was the moment – and as a result we actually have a photo of the proposal itself. You know the way it is; my keys are always in the last pocket I check. It took me a couple of seconds to find the box with the ring. My mind was a mess: *How am I going to get the ring out, and not drop it in these hurricane winds?*'

What did he actually say? I want to know.

The couple are silent for a moment and then burst into laughter.

'I missed the question completely!' admits Meghan. 'The wind was so loud. And he was holding the ring box so tight with both hands that I couldn't see what it was.'

'Well, I didn't want to drop it! And there may have been a lump in my throat,' Paul adds generously. 'But there was definitely a question in there. You took ages to answer!'

'I didn't want to say anything because I didn't want to make a fool of myself,' she counters. 'Until he finally asked: "Are you going to say yes?"'

Foolishly perhaps, Paul had assumed that the ring was going to stay in the box until the couple were safely back on solid ground, but Meghan was having none of it, sliding the ring straight onto her finger and dancing along the platform. 'I'd actually really like to go up the bridge again, because after that, I spent the whole time looking at my hand!' she laughs. 'But it's never going to top the 25th of September 2022! Nothing will.'

Once the group had descended, staff lined the exit in a guard of honour. 'The staff made it really special; they were just so genuinely pleased.' Paul grins. 'Imagine it: in the portacabin, they tell you that the winds are too high, and the rain is too heavy and that it may not happen. Then you come down again, and you're engaged, and the sun is almost shining. It was just a brilliant, brilliant day!'

MODEL MAKER
Michael Dineen

'MY ASPIRATION IS to see Scotland on the Lego shelves. The company have a range of amazing landmarks from around the world – recently they even added a huge set to build the Eiffel Tower – but there is nothing in the Lego range representing Scotland at the moment. As a Lego enthusiast and civil engineer, I thought I might try to change that.'

As you do...

A STEM ambassador, Michel Dineen feels strongly about inspiring the next generation of engineers. When he was seconded for a time to Transport Scotland, consequently spending a lot of time at South Queensferry, the idea of a Lego model of the Forth Bridge took hold. 'It's so good for teaching STEM concepts! A Lego model would tick all of the school boxes: transfer of forces, tension and compression, it's all there. Even as I was building my own Lego model to submit, I learned that the principles were all the same. When you look back, the Victorians built the towers first, and then constructed the cantilevers out from there evenly so that the towers would stay balanced. The same applies in Lego. If you don't build out evenly, your tower will fall over.'

The route for submitting an idea to the company for consideration is complex. First, the model must be built. Then it must be photographed with strict criteria, and uploaded to the company's *Ideas Platform* where it must garner ten thousand votes of support before being considered for inclusion in the Lego range of products.

Michael's first step was to approach a reputable freelance Lego designer with a view to commission a Forth Bridge Lego model. The freelance quoted an eye-watering fee of £10,000 and Michael decided to set to work himself. 'I thought, let's investigate the feasibility of doing this. It hadn't been done; I knew that much. It took me four months in total: three months of finding out what *didn't* work, and then a month of building my model once I figured out what *did* work,' he admits cheerfully, 'a couple of hours every night.'

'Everybody thinks that I am this Lego nut with drawers full of Lego – that's not the case; I don't have lots. I had to go online and order from the company website. For three months, I spent

a fortune on pieces I *thought* I needed!' he tells me. 'The A-frame towers were challenging – the spacing of the studs on the individual pieces restricts what you can do. It was trial and error. On one occasion I went to the Lego Store in Buchanan Galleries for a bit of inspiration and saw a set box which contained the precise piece I needed at the time – a ball and hinge joint. I don't even remember what the box was for – I bought it for the one piece.'

Due to the extremely symmetrical nature of the Forth Bridge, finishing the model was less tricky. Michael explains: 'The Forth Bridge consists of three towers, and then 12 arms form the cantilevers. Once you have built a complete tower, you just buy a ton of Lego and repeat the process. The connecting girders presented no challenge really. However, I did have to individually paint 240 little bridge deck parts by hand – the pieces I needed only came in white!' In a surprise twist, Michael's Forth Bridge outgrew even the longest room in his house at 4.7 metres in length. He had to take it outside to take the all-important photographs for submission on Lego's *Ideas Platform*, with white sheets draped over the back fence to form the obligatory white background. 'I tidied the images up in Photoshop, uploaded them and waited for verification. It was nerve-racking. After two hours it was rejected, but I realised my mistake: I had selected a picture of the *whole* bridge instead of showing what a customer would get in a box. Due to the size of the model and the maximum number of pieces allowed per box, I was proposing a set which would contain only one-third of the bridge. If you wanted the entire bridge, you just needed to buy three boxes.'

Did he ever consider simply building a smaller model?

He sighs regretfully. 'But then you lose all the detail…'

He uploaded a more accurate image which was quickly verified. However, the next challenge was garnering the necessary votes to get the design assessed. 'Lego's *Ideas Platform* gets around 50 submissions a day. There is some amazing stuff and some car crashes on there,' Michael adds, 'and you can vote for the ones you like. Essentially, I was now looking for ten thousand people to say: "I'd buy that."'

In an attempt to boost the hundreds of votes he had accumulated by promoting his idea on social media, Michael took his entire model of the Forth Bridge to Glasgow's Central Station for two days and was featured prominently in an episode of the BBC documentary series *Inside Central Station*. Finally, two years after he embarked on his mission, Michael secured his ten thousand votes of support.

'Imagine it like a holding room – all the ideas with the necessary votes get through the door, and every 12 weeks or so, a panel comes to assess the models in there. Normally there would be seven or eight in the room. When mine got assessed in the middle of the pandemic, there were 57 in there. Everyone was at home, lots of people were building and lots of people were voting. After a few more agonising weeks, the dreaded email arrived, thanking me for my efforts but advising that my journey had reached the end of the line.'

When Michael asked for feedback, the company declined. 'They said they couldn't comment as it was "commercially sensitive". It's a cop-out if you ask me. But now the model is displayed in the education centre by the bridges and youngsters who visit can see it up close.'

I am amazed at Michael's dedication. But I'm also curious: Does he remember how many pieces are in his bridge?

He hesitates. 'I do. But Transport Scotland run a competition to guess the number of pieces from time to time. I can't give you the answer here now, can I?'

NETWORK RAIL ASSET ENGINEER
Jamie McLaren

'THE BREAD AND butter of my job is protecting the Forth Bridge, and other bridges, by evaluating its condition and prioritising repairs. I am an asset engineer.'

I am not about to let on to Jamie McLaren that I have never heard of an asset engineer, so I ask him to clarify. Does he deal with budgets? Is he a number cruncher?

He laughs. 'No, no, I leave that to people who are good at that sort of thing.'

Thankfully, he elaborates before I have to admit my ignorance.

'Bridge examiners go out to the Forth Bridge and produce Network Rail a condition report. That goes to me. I look at the findings and with my team, we evaluate and prioritise repairs, whether they be critical or not so pressing. I then take all of that to the real number crunchers, present the facts, and they let me know what can be done. It's simple, but it's teamwork.'

It sounds clear enough when he puts it like that. Jamie has been involved with the Forth Bridge's maintenance for Network Rail since 2014. 'We generally work in control periods of five years, and I came in about midway through control period five (CP5). Of course, all of our work is ongoing and constantly re-evaluated, but every five years the structures department set out a proper business plan. At the time, no dedicated person was specifically focused on the huge bridges in the east, the Forth Bridge and the Tay Bridge. People had bits of knowledge, but there wasn't one go-to person when people moved on or retired – not until I came in. It was all new to me and it took a lot of research – what happened when, what is ongoing. I am still learning on the job.'

I briefly met McLaren on the very first day of my Forth Bridge residency, clad in Network Rail high-visibility vest and hard hat, like most of his colleagues. His job used to be based at the Forth Bridge, but he now works out of Network Rail's head office in Glasgow, with perhaps two or three site visits a month. He explains: 'I do liaise closely with the project team on the bridge, and also with our contractors Balfour Beatty, but my remit has widened now, so basing myself in Glasgow made sense. However, we do have a lot of communication between everyone involved, including a monthly

progress meeting with the Forth Bridge guys.'

So, what sort of work might he request? As a non-engineer, I am interested. 'Well, we deal with all sorts of repairs, ranging from something that needs immediate attention to things that can wait for even ten or 20 years,' he begins. 'It's all about managing risk, looking at the defects and using engineering knowledge of how it may deteriorate over time. But yeah, some things are urgent, and need done ASAP. If someone spots a high-risk defect, they call Network Rail who pass it to me. I'll instruct repairs if necessary. As I said, it's all about degrees of risk.'

I am struck by the responsibility on this young man's shoulders, day in and day out. But perhaps it should come as no surprise – just as Jamie now protects the Forth Bridge for future generations, so his grandfather did during the war. Maybe Forth Bridge protection is a family trade. He is thoughtful.

'I suppose you could put it like that, yes. My grandad worked on the battery placement on the Dalmeny side during World War Two, to defend the bridge. He has always had a great affinity with the area for that reason. He also used to work in a steel foundry, so he had a real appreciation, a fondness and an understanding of how this structure was made.'

I am moved at the way Jamie has picked up his grandfather's baton as a protector of this crucial cog in Scotland's infrastructure. He continues: 'I took my grandparents up the bridge once, right to the top. As part of my job, I used to be a bit of a Forth Bridge tour guide, showing groups around. They were mostly internal parties – Network Rail team away days, graduate scheme groups – because let's face it, the Forth Bridge is a great selling point. Taking my grandparents along was personal though. They joined a small group – I still remember it, mainly because of the weather conditions which I had never encountered before or have since. When we first reached the top, visibility was almost zero, but during our time up there, the fog cleared, revealing the river and all the views around; it was just magical. I must have been up there hundreds of times since, but I have never experienced a dramatic reveal like that again, and my grandparents were there to see it.'

The Forth Bridge may be a triumph of Victorian engineering, but it needs ongoing maintenance. Is he concerned for the future of the bridge at all?

This time there is no hesitation.

'Concerned? Certainly not, it's in good condition, and I am part of a great team committed to looking after it. All of us feel great pride in what we do, helping to maintain it. Not everyone gets to do what we do. If we make progress or fix something on the bridge, I can sit back and think "I've been in charge. I did that. I brought that to the table." I and those around me protect and maintain the bridge for this age and the next. And there is great satisfaction in that.'

OUTDOOR SWIMMER
Gina Bees

'WHAT DO WE do? Well, we basically throw ourselves into the freezing Forth, all year round. Let me tell you, it's more pleasant now that it's warmer and we are not freezing our bits off!' Gina Bees is a breath of fresh air – curly-haired, smiley-faced, bespectacled and congenial. I warm to her immediately.

She first gave outdoor swimming a try during the pandemic. 'It was during lockdown. I met one friend, which was all we were allowed at the time – and that would be my social contact for the day. When restrictions eased, I joined existing groups at Wardie or Portobello. But then I asked myself: why am I driving 45 minutes or longer to jump in the water when I could do that here? Why don't we have a group here in Queensferry?' The Mental Health Swims movement aligned closely with her own vision for such a group. 'Many people feel daunted about doing this kind of thing. They don't know what to expect; they don't know what to bring; they don't know who else will be there. We can give people the information they need in advance and create a supportive, welcoming environment. It's community over competition; dips, not distance. If people want to train for a biathlon, there are other groups for that. We welcome anyone who wants to give this a go.'

Gina, originally from Dorset, has lived in South Queensferry for the past decade. Based in London, Singapore and Edinburgh along the way, she describes the town as the home she didn't know she'd been looking for. An employee of Scottish Water, her day job mirrors her favourite hobby. 'There is definitely a water thing going on – swimming is the thing. I do love kayaking and paddle-boarding, but swimming – physically being in the water – puts a different perspective on the world.'

She prefers to call it 'outdoor' rather than 'wild' swimming. 'I don't think of it as wild, no. That word could sound intimidating, and we are not about that. There is such a benefit to being outdoors in the water. We all live in our heads too much, don't we, with our busy lives? Well, when you swim, for the first minute or so your mind empties entirely – all you can think of is how cold it is, how it stings. And then you slowly reconnect with your own body and that zingy feeling begins – and you realise that you can

do hard things. Bobbing around in the bracing water helps regain perspective on whatever else is going on in life. Even the Forth Bridge looks different from the water. It's always a stunning view of course, but in some ways it's an anchor too. You know exactly where you are, and that's reassuring.'

The sense of community Gina has fostered is the secret to the group's success – in its first year, regular attendance has soared from two to an average of ten to fifteen swimmers with a lively Facebook group to boot. Gina personally bakes for each of the group's swims, and there is much laughter while members chat over medicinal tea and cake. 'Oh, the stories I could tell you,' Gina begins with glee. 'The beach we use is also frequented by dog walkers. Well, a dog ran up to my knickers and bounded off with them. The owner was mortified and shouting at his dog, but I realised that on that occasion I'd have to go commando on the way home. I always take real care over securing the tin of cake while I'm in the water. If only I had taken the same precautions with my underwear!' she giggles. 'We do have a laugh! Over the months we have had all sorts of people: locals and incomers, Queensferry folk and others from a wide catchment area. Some wear wetsuits and some don't; some come once, and others appear without fail.'

One of the logistical challenges of running the Mental Health Swim is that swimming in South Queensferry is entirely tide dependent. It means that the group cannot meet at a regular time. 'If you turn up and the tide is out, you are wading knee-deep in mud,' Gina explains. 'No one wants to do that! So, we let Mother Nature set the schedule. Most of our swimmers prefer to meet in the morning so that it doesn't impact too much on the rest of the day.'

One particular memory stands out for her. 'A lady came along for the first time. She was clearly nervous, not sure if she should be there. She was a bit self-conscious about getting undressed on the beach, but once she realised that we were not a bunch of lithe elite athletes but all sorts, she went ahead. We walked her in. Once in the water, she had a little cry. I'm welling up just thinking about it. "I didn't think I could do it," she said. That was a powerful moment for me too. These moments make it all worthwhile.'

Hearing her passion makes me want to run home and scour the internet for oversized wetsuits. However, I do have one final question: How clean is the Forth?

She chuckles. 'Well, put it this way: personally, I never stick my head under! At your own risk, people! At your own risk!'

PRINCIPAL TRANSPORT CURATOR
Louise Innes

MADE FROM GIRDERS: OUR FORTH BRIDGE

'I HATE YOU, Louise! That's my favourite bridge!' exclaimed a friend and colleague when hearing of the place Louise Innes calls home: a house high on the hill in North Queensferry, with floor-to-ceiling windows. The entire view is filled by the Forth Bridge.

On a clear day, Louise can observe a train leaving Dalmeny station on the other side of the water, leave the house and still make it to nearby North Queensferry station in time to catch it. On the day we speak, however, all is obscured by sea haar. 'Today I can't see the bridge at all. That's one of the amazing things about it,' she enthuses. 'You can sit here in our snug with a glass of wine, and the Forth Bridge's colour changes constantly, depending on the light on the water – sometimes it's all but invisible, and sometimes it's jaw-droppingly spectacular, especially at sunset. I never tire of it. Never.'

Born and bred in Edinburgh, Louise Innes remembers throwing pennies from the train windows with her father. 'And there was a Broons cartoon too, do you remember that? A cartoon where they went fishing beneath the Forth Bridge but came back with lots of pennies instead.' I do not, but I can picture it.

More recently, however, Louise Innes experienced the bridge as a commuter to her job as Principal Curator of National Museums Scotland's Transport and Industry collection. 'For the last 12 years I crossed the Forth Bridge at least twice a day by train to get to central Edinburgh. It is such a vital, vital link in the whole of the British railway network,' she emphasises. 'On one occasion I went across six times in a single day – it's just so easy. With my season ticket, I could head home, get changed, and head out again. The Forth Bridge was my daily companion then, and it still is.' Now retired, Louise is a keen amateur sportswoman and has rowed beneath the bridge hundreds of times. 'It never fails to excite me, the view upwards through the steel work.'

Louise decided that she wanted to return to Scotland while working as a map surveyor for the Ordnance Survey. With no prospect of a transfer north, she handed in her notice and moved. 'It's interesting because I moved into museum work sideways. I had never envisioned myself in that job. But once I was there, I spent

the rest of my life there, finally as Principal Curator. The world of heritage and museums has changed enormously. Now many of my colleagues come straight out of university, and there are so many women working in technology and transport heritage. That was definitely not the case when I started.'

In the National Museum of Scotland, I looked after the collection on land, sea and air transport which latterly included bridges. On the first-floor gallery, we have pieces of the collapsed Tay Bridge for example, and we featured a paint mixer used on the Forth Bridge in an exhibition – you know how iconic painting the Forth Bridge has become. And when the Queensferry Crossing was built, we collected some of the tubing which encased the cables for our collection too.'

However, the most memorable Forth Bridge acquisition Louise made for the museum was an original oil painting which became an iconic railway poster, painted for British Railways (BR) in 1952 and showing the Gresley A4 Class Pacific Plover locomotive crossing the bridge. Louise laughs. 'The artist Terence Cuneo had sketched it on the day the King died in February 1952, in gales of over 50mph from a girder high above the track. Apparently, he wore a flying suit, duffel coat, balaclava and mittens as well as long woollen underwear – and he was still frozen. I love that the painting shows the bridge as a working girl, not just a pretty face. The image is just iconic. It formed part of an exhibition called See Scotland by Train.'

Retired since just before the pandemic, Louise is struck by the way her home village is inextricably linked with the bridge which towers above it. 'North Queensferry is a small village of just over a thousand people, but we do get a lot of Japanese visitors because of Kaichi Watanabe, one of the engineers who worked on the Forth Bridge construction. I think he is a lot more famous in Japan than he is here. He was the one perching in the middle in the famous human cantilever photograph.'

Can she think of any downside to living so close to the Forth Bridge, given that her professional life also touched on the world-famous structure?

She doesn't hesitate. 'No. I love living here. The trains don't disturb us at all, and you get the best views here. The first Sunday morning train is the one for Aberdeen, at 8.20am, and 20 minutes later the first southbound one to Edinburgh goes past. The trains just become part of your body clock. The bridge is part of life, part of all life. It's interesting, always.' She pauses.

Then she repeats: 'No. I have nothing but happy memories of the Forth Bridge.'

QUEENSFERRY COMMUNITY STALWARTS
Karen MacGregor and Terry Airlie

AS THE SCHOOL symbol, the Forth Bridge appears on the badge of every item of school uniform at Queensferry Primary School. With 535 pupils, that's a lot of bridges – a daily reminder to Head Teacher Karen MacGregor of her school's connection to the famous landmark. 'It also appears on our letterheads and documentation – even our school values are accompanied by a child's painting of the bridge. The canvas of that image is literally the first thing that people see when they enter the building,' she announces with pride.

Karen has been Head Teacher of Queensferry Primary School since 2010, having previously taught there 1997–2002 in less senior roles. What strikes me most is her undisguised enthusiasm at being approached for an interview. 'I LOVE our bridge – it has a very special place in my heart and across our school,' she replied to my initial email. 'I would be happy to be interviewed.'

A mere day later she had carved out a slot for me in the busy run-up to the end of term.

'It's really important to us that the children learn about the bridge,' she begins. 'Honestly? I think all Queensferry children believe that bridge really does belong to them. I'll give you an example: in the run-up to the millennium there was a countdown clock on the bridge, do you remember? We took some of our pupils to Glasgow on a trip around that time. When we returned on the bus and the bridge came into view, the children spotted that the clock had now lit up and they burst into spontaneous, completely unprompted cheers. They feel that the bridge is everything to do with them.' Young learners at the school study the bridges in Primary 3 and return to their local heritage in Primary 6 when they dress up as characters from Queensferry's past and complete a heritage trail around the town.

Even to the teachers, the vistas of the Forth Bridge are a constant reminder of what they are trying to do in inspiring the next generation. Karen uses the bridge as a metaphor, a visual aid for the school's curriculum strategy: 'By likening the principles that underpin our curriculum to the spans, and the curriculum design to the tracks which would take pupils forward towards their next destination, we can visualise what we are trying to do in a way

that is unique to our school. It just feels like the natural way to express it,' she explains.

But for her, the Forth Bridge carries personal significance, too. Originally from Fife, Karen now lives in Edinburgh – at every stage of her life, crossing the bridge has been synonymous with home. A family photograph with a Forth Bridge backdrop now graces her desk alongside another, even more prized picture. 'That one is a tiny black-and-white photograph, taken by my papa, perhaps as early as the 1930s or '40s. It's just the bridge, taken from the quarry, but that picture has sat on every single one of my desks as a teacher.'

The community's immense pride in the structure is evident in her voice. 'We often take the Forth Bridge for granted and underestimate its significance because it is on our doorstep. But when the bridge was granted UNESCO World Heritage status, there was such a sense of joy, such pride! It's a global icon. In fact, a pupil in my class went overseas for the holidays over the turn of the millennium. It was a transatlantic trip, but she returned with the newspaper of the 1st of January. The front page showed celebrations around the world. And there it was, a picture of *our* bridge, right beside Sydney Harbour Bridge and all these other famous places. And in this job, I can see how much it is loved.'

Queensferry and District Community Council secretary Terry Airlie remembers Karen as a teacher, back when his own children attended Queensferry Primary School. His daughter completed the Heritage Trail she spoke of as a Primary 6 pupil in the '90s, and his oldest grandchild has just repeated that experience. 'We moved here as a family in 1988, just in time to witness all the centenary celebrations in 1990,' he tells me. 'My kids were brought up and educated here. I suppose I joined the Community Council to give something back to the place.' He has now been a member for almost 20 years. The Queensferry and District Community Council exists to represent the interests of local residents to the City of Edinburgh Council and to the Scottish Government. 'Much of our business is done by email, but we have monthly meetings, nine elected members and some nominated representatives of local interest groups too,' Terry explains.

He is evidently a man who is comfortable with detail – his answers are considered and precise and my pen is flying to keep up with him. What Forth Bridge-related business has the council been involved in? What types of things are you pushing for? I ask.

'Well, we have some tie-ins with Network Rail, particularly as regards the UNESCO World Heritage nomination and inscription. We had perhaps expected some more tangible benefits of the UNESCO success by now: improvements to road services, infrastructure and public transport and so on – it is a frustration. Or take the Forth Bridge Visitor Experience – we were made aware of these proposals at an early stage. There were extensive plans for both sides of the Forth, but it has gone a bit quiet. I suspect funding is an issue; that would be typical.' His honesty is refreshing.

In his role with the QDCC, Terry has met a range of interesting individuals. On one occasion, he found himself sitting beside the then First Minister on the *Maid of the Forth* as it crossed the water to mark the end of the major Forth Bridge refurbishment. 'Alex Salmond was talking away to me, about the day, the blustery weather. The thing is, he kept calling me by name. I puzzled – how did he know who I was? It was only later that I realised I was wearing a massive name badge saying *Terry*!' He laughs.

One of the Community Council expeditions he recalls is visiting the Scottish Parliament to share some local concerns about the proposed Queensferry Crossing. 'When a government plans to undertake the biggest construction project since the Second World War in your backyard, you need to make your voice heard,' he quips. 'We needed to know what the tangible benefits were going to be for the local community, how traffic was going to be realigned and so on. But actually, being involved in the discussions about the Queensferry Crossing led to some great learning about the original bridge: the Forth Bridge construction process, laying the caissons and so on. It's perhaps a bit over-engineered, but what a marvellous structure!'

I am interested in his perspective on the future in this community he represents. What would he like to see happen? He doesn't hesitate. 'Two things: I'd like to see the lovely museum in South

Queensferry re-open. It would be perfectly placed to tell the story of the bridge. We have a world-class attraction, over 30 cruise liners a year dock right here, their passengers walk through our streets – and the museum is closed! Secondly, I would like the Forth Bridge Visitor Experience to happen, and soon. I have crossed the Golden Gate Bridge, walked across Brooklyn Bridge and seen a lot of other World Heritage sites. We should have a world-class visitor experience to offer here, there is no doubt about that.'

ROYAL NATIONAL LIFEBOAT INSTITUTION
CREW MEMBER

Julie Dominguez

'THERE IS NO science to it. Suddenly there is a racket coming from this little thing in your pocket, your pager. It sounds old-fashioned, but believe me, it works better than any phone. Queensferry is a bit of a black spot for phone reception.' Julie Dominguez shrugs. 'Then you drop everything and go.'

As a hands-on operational lifeboat crew member at South Queensferry RNLI, Julie is ideally placed to give me a first-hand insight into the world of these volunteers.

She explains the process: 'As soon as your pager bleeps, I need to get myself down to the lifeboat station as fast as I can. That can be frustrating. We are all volunteers, so we could be stuck in traffic, with the people in front of us oblivious to the fact that we are trying to respond to an emergency. You have to stay calm and safe, despite the fact that the adrenaline is going – someone is in trouble out there and needs help! We are not told what the call is about. As soon as we get to the station, we make for the changing room and get our kit on as fast as we can. Only then do we hear what the callout is about and some of us can begin planning how to address that situation, while others get the boat ready. We set up aerials to send and receive signals, and the radar goes up too, just so that we can see and communicate with others on the water.'

Julie Dominguez has just passed her helm exam, so congratulations are definitely in order. In lifeboat circles, the volunteers gain qualifications called 'tickets' and she has just earned the right to shoulder more responsibility. 'There isn't really a hierarchy in the team, although there are different roles. We are all one family, and we have each other's back regardless of what happens.'

Julie began volunteering with the South Queensferry RNLI around five years ago. Her job at the time was in a sales office with an incredible view over the river, watching ships go up and down the Forth and observing the progress of the Queensferry Crossing construction. 'I wasn't actually living in the Ferry at the time, but I was working here all week, so I was allowed to join the crew,' Julie explains. 'All my employers have been really supportive of my lifeboat volunteering and have allowed me to work the time back. My degree was in Leisure Management, and I thought some

of my skills might be useful to the RNLI. Now I am not in sales anymore; in my new day job I work on the water too, next door to the station in fact.'

So, what type of situations does the lifeboat encounter in the waters beneath the Forth Bridge? She pauses for a moment. 'There is such a variety, you just never know what you are going to get. But I do quite like the satisfaction of towing a struggling vessel back to safety. When we get a shout like that, we don't know if there is damage to the vessel, or if anyone is injured – we just go. You can see how easily it can happen: people might be on a sailboat at Port Edgar Marina, and on a beautiful sunny day with a westerly wind, they set out. But by the time they try to get back, the wind has died down and their engine doesn't work. That's when their nightmare starts as they drift further and further out to sea. As soon as we approach, they know that they are going to get home – even the sight of our orange lifeboat calms them down. We rig up the tow and one of our crew joins them on board, to reassure them. You can literally see the stress go from their faces, and that is so satisfying.'

On another occasion, Julie's team received a shout during an open water swimming event from shore to shore. As a novice open water swimmer herself, Julie can identify with the swimmers' panic. 'It was a mass swim. Something must have changed; something must have happened. I don't know what, but some swimmers got tired earlier than they perhaps expected, and they began to struggle. There were just so many of them, so it was quite a stressful situation for us too.' Julie physically gestures as she describes hauling the tired competitors into the rescue boat and I am struck by the physicality of her role. 'Once they were safe you could see the relief in their eyes,' she concludes.

Occasionally, the team encounter humorous situations too, and not all of their emergencies are human. 'We have rescued a few dogs along the coast at Gypsy Brae, for example. The dogs jump over the wall, not realising the steep rocks on the other side. The animals end up getting stuck and we need to get them out. But one of the most amusing callouts was a train passenger crossing the Forth Bridge who was convinced they had spotted a drowning person in the water. It was done with perfectly good intent, but it was actually a cone-shaped buoy with a stick protruding – from a distance it may well have looked like a waving arm.'

Julie clearly values the RNLI community around her and lights up when she talks about her fellow volunteers. 'It feels like I have a family around me. Whether it is at training or during a shout – honestly, sometimes I can't believe how well we all get on.'

Given the stunning location of the RNLI station, does she ever take her surroundings for granted? I wonder. 'No,' she says with a calm smile, 'I never tire of the views. Even when we're speeding across the water during a callout and we're really busy, I do try to take a second if I can – to look up.'

SKIPPER
Scott Aston

MADE FROM GIRDERS: OUR FORTH BRIDGE

SCOTT ASTON REMEMBERS first passing under the girders of the Forth Bridge in his father's old fishing boat as a boy. 'I must have been about ten years old I reckon. We grew up in West Lothian, not quite within sight of the bridge but close, and my dad had always been into boats. I remember him telling me to watch out for trains, and like everyone else, I thought that the trains ran over the top of the bridge, roller-coaster style. We recently put a post on our *Maid of the Forth* Facebook page asking our followers: Hands up anyone else who believed that as a kid. One commented: "I am not going to disclose the age when I realised that the trains ran through the towers of the bridge rather than over." It's funny!'

Scott is the skipper of the well-known vessel which takes visitors under the bridges and to Inchcolm Island, with plenty of wildlife watching opportunities on the way. His family have run the business for three decades. 'My dad bought the business in '93, and I worked on the boat as a crewman when I could. But I wanted to go to university – I had always loved maths and physics, so I was off to Glasgow University to study Civil Engineering.' He laughs. 'I guess that is bridge-related too. However, after eight or nine years of working as a civil engineer, I got bored. I was either going to work for another company or do something different altogether. That's when my dad suggested coming back as a skipper, and I did. I've been doing that for the last 15 years or so. My brother-in-law is the other skipper, so it's very much a family affair.'

I ask him to explain the process to me as a non-sailor – what is involved in skippering the *Maid of the Forth*? 'Well, as soon as we leave the pier, I check the traffic comms and speak to Forth Navigation. I may need to change route, depending on who else is out there on the water. It's like crossing the road in slow motion. If a big ship is coming up the Forth at 16 knots while I am trying to steer mine across its path at 8 knots, it can sneak up quickly. I am basically trying to avoid a slow-motion crash. Thankfully, they will put local pilots on any incoming ship – someone who actually knows these waters. I will speak to that pilot to agree whether I will cross in front of them or behind them, to make sure everyone stays safe – there is always a lot of communication. And yes, going

backwards and forwards to the island can be monotonous, but the docking is never the same from one day to the next. It can be challenging; you always have to pay attention to the changing winds and weather, and to the swell of the sea. From time to time, rough weather makes life interesting for us,' he adds, grinning. The *Maid of the Forth* provides a shuttle service to the much-visited Inchcolm Island, site of a historic abbey and fortifications dating to both World Wars, as well as beaches and seabird colonies. Scott explains: 'We drop one group off and return to pick them up, dropping the next group at the same time, if that makes sense.'

Despite being responsible for safety and navigation, Scott does have a fair amount of face-to-face contact with passengers, too. 'You're there to greet them as they board the ship, and people always like to chat to the skipper, don't they? We usually let the kids in to see the wheelhouse and touch the steering wheel at some point.'

On the way back, the *Maid of the Forth* passes under all three bridges, but Scott is far from ambivalent. 'I go under the Forth Bridge so often, but I still look up. It is immense. I travel under there eight times a day, and every single time I can spot something new. The old toilet cubicles, or the tea rooms for the workers by the track. The slight bends and dents in the metal. And there are all these ladders that you don't really notice from land – I have nightmares about being lost on the Forth Bridge from time to time, and I blame the labyrinth of ladders for that. Seriously though, every time I think about how that bridge was built in Victorian times, I am gobsmacked all over again. Seriously impressive. I can't get over that the engineers had to wait six months for the right temperature before the last pieces could go in. They couldn't afford to cut corners and they needed to finish that job properly.'

In addition to the main boat which offers visitors a pre-recorded commentary, the family also run trips in a small rib for groups of up to 12 passengers, with live commentary. 'I enjoy doing that – over the years you learn so much and I don't mind sharing it. You always think you know it all, but occasionally the passengers also teach me something new. On one boat trip, a passenger said: "See that steel tube up there? It's a 12-foot diameter steel tube, exactly like the London Underground. Same designers, Benjamin Baker and John Fowler." I went home and googled it, and it's actually true!'

Scott meets many memorable passengers, and their questions can be entertaining. He recalls: 'One passenger who had literally just stepped off a huge cruise liner asked me if the River Forth connected with the sea – I wanted to tell him that his very presence was proof that it did. Others ask if the Forth Bridge swings open at all,' he laughs. 'But they all notice when the skipper of the boat steps out of the wheelhouse to take a photo, and I still do that. The best view is early in the morning when we bring the ship over from Port Edgar Marina and you see the sun rising behind the Forth Bridge. I can tell you now, I never get sick of that. Never.'

TRAIN DRIVER
Gavin Black

'THE FIRST TIME I drove over the bridge on my own, I thought: This is it. I've made it.'

Gavin Black had always harboured a secret ambition to drive trains. 'I mean, come on! Everyone wants to be a train driver! ScotRail recently advertised for the job and 20,000 people applied. Twenty thousand! It's a really sought-after position.' In some ways the bridge was familiar territory already. The young Gavin attended school in South Queensferry and was often distracted by the view from the windows on the top floor. On one occasion a teacher he calls Captain Caveman berated him for his lack of focus. 'Will ye stop swinging on that chair and staring out the window! You'll never get anywhere staring out of a window!'

'To think that I now make a living staring out of a window! I do love a good view,' Gavin laughs. 'And I have loved trains from a very young age. I think it's laid into you if you grow up around the Ferry, there is such a pride in the bridge in the community. I joined my father's business as a delivery driver, crossing the Forth Road Bridge daily, and looking across at the trains on the Forth Bridge. It took my wife to kick my arse to finally do something about it. Getting into train driving is very competitive, so we decided to travel to Doncaster where you could sit the psychometric test privately – that way I had independent proof that I was suitable for the training. When you think of it, every new driver is quite an investment on the part of ScotRail – the training can take up to two years. With this test under my belt, I was able to skip a whole load of steps, and that got me through the door.'

Now he travels across the Forth Bridge by train up to 30 times a week. What is that like? What does he see?

'For a start, driving a train across and travelling as a passenger are completely different things. As a passenger, you can only see where you are. As a driver you can see where you are going! And once you're that close up, you can see everything around the track. You see how the structure expands in summer and you see that it's far from straight – it zigzags around everywhere! And you are so close to the edge, especially on the viaducts; there is a proper sense of height. You'd think it would open up as you go onto the bridge.

It actually closes in as you enter the structure, but I love looking at all that too. I grew up in a family of engineers, so I do tend to notice the small things. I always ask myself: why is that there?'

Gavin recalls visiting an exhibition of Forth Bridge memorabilia at a local hotel. 'There was a massive bolt there, colossal. It must have been half a foot thick, with a two-foot head, something like that. I found out that four of these things hold the suspended bits, the central girders, in place. And once when I drove a train across, I saw just such a bolt head, and I thought: *Ah, that's where that goes!*'

However, not all journeys go smoothly. During a recent storm, he felt the whole train sway, churning his stomach, and there are often dead birds by the track. One particular incident has stayed in his memory: 'It was during the refurbishment, so around 2008 or 2009. I was driving a two-car 158 Diesel Multiple Unit across the bridge. When I drew into platform 13 at Waverley Station, the Driver Team Manager was waiting for me. I thought, *oh my goodness, what have I done?* If a manager is waiting for you at the end of the platform, you automatically think the worst. But no: apparently, someone had dropped a scaffold pole from the bridge onto my train from above. It fell onto the rear carriage, so I hadn't heard anything at all. "Well, we need a report," the managers told me. "Just write what you know." I had to fill it in honestly, so I literally wrote "I know nothing". They take this sort of thing really seriously; the fireworks kick off when stuff like that happens.'

As a schoolboy, Gavin often bought chips by the Hawes Pier during school lunchtimes and discovered a profitable pastime. 'In the ferns by Jacob's Ladder, the steep steps behind the Hawes Inn, right beneath the access track under the viaduct, I found a broken rivet from the bridge. It still had a bit of the orange paint on it. I sold it to an American tourist for a tenner – that was a lot of money to a boy in the eighties. I never told my pals though, otherwise they might have tried to look for rivets too. Over the years I found three more, selling them all to tourists. I regret that now, but if you are a kid, all you can see is the money.'

There are plenty more memories of growing up near the bridge: a school assembly attempt to recreate the famous human cantilever image, with nearly disastrous results. Playing in the school rock band for the bridge's centenary celebrations. Appearing in the background of an Alan Titchmarsh live broadcast for the BBC. 'I was in my final year, and one of the few pupils at my school who had a car. I drove into the Ferry to watch the events. But the whole Binks car park was full of BBC lorries. Some people told me about the live broadcast. If you watch it on YouTube now, you can see me in the distance at the back of the harbour, carelessly trying to get on the telly.'

Gavin only flounders a little when I ask him for his favourite weather conditions on the bridge. 'Oh, when it's foggy. Not all the floodlights are switched on now, but they have moved them to track level, and in dense fog they light up the air above them – it's really eerie and quite magical. But no, maybe early morning sunlight, looking across the estuary, that's fantastic. I remember the first time I drove onto the bridge in a High-Speed train, early in the morning, in this massive locomotive. When you're coming onto the structure at 50 miles per hour, the ground around you just falls away. All of a sudden, you're 150 feet in the air! It's a thrill. There are five million people in Scotland, but I am one of very few who are lucky enough to see this. Every time I pass over the Ferry I reminisce about my youth, hanging out with my mates all parked up in our Ford Capris along the Hawes. We were all down there looking up. Now I find myself up here looking down and wondering how I got here.'

UNESCO WORLD HERITAGE LEAD
Miles Oglethorpe

MADE FROM GIRDERS: OUR FORTH BRIDGE

'IT'S A GIRL,' Cabinet Minister Fiona Hyslop announced, standing atop the Forth Bridge with Historic Scotland's Head of Industrial Heritage, Dr Miles Oglethorpe, in December 2013.

'Excuse me?' asked Oglethorpe.

'The bridge is a girl! She is far too curvy to be a boy.'

It is Oglethorpe's first response as I ask him for memorable Forth Bridge moments – and he has had a lot, having first encountered the structure professionally in 1989. At the time, he was working for the Royal Commission on the Ancient and Historical Monuments of Scotland, and the project was to photograph the Forth Bridge for its centenary in 1990. 'There was no hoist then,' Miles Oglethorpe begins. 'We had to walk inside one of the top girders from the North Queensferry end – it was a really intimate, rivet-by-rivet encounter with the bridge. You notice new things then – how the whole bridge tapers like a letter 'A' when you look at it end-on. As you ascend within the girders from the stone portals the space inside the girders is small, so you have to double up and conditions are very cramped – but they taper out. By the time you reach the top they are almost double your height and you have to climb out – it's weird!' On that first visit, he saw many of the bridge's lesser-known areas first-hand and close up: the wooden walkways, some of which were worse for wear, and the peeling paint which, contrary to popular perception, was not down to rust, but simply due to the Victorian practice of painting over old paint, resulting in layers breaking off. 'And there was an old Forth Bridge worker's toilet at track level,' he adds, 'literally a ramshackle hut offset from the track. It just hangs over the water; I can't even remember if there was a door or not. The seat is just a plank with a hole in it. All I could think was: What about strong, unexpected gusts of wind?' He laughs heartily. 'Let's just say that made an impression!'

More than three decades later, Miles Oglethorpe's role is far removed from this tactile first encounter. Despite his wider responsibility for Scotland's industrial heritage, Oglethorpe also serves as chair of the Forth Bridges Forum's World Heritage Management Group, leading on the process which resulted in UNESCO World Heritage listing for the iconic structure. He admits:

'I'd say Forth Bridge work is still taking up much of my typical day, liaising with our many partner organisations, including Transport Scotland who I think of as the mother ship. Owners Network Rail and VisitScotland have also been amazing, as have the local councils and community groups. The thing is, working with many others in this way is really educating – all have their own perspective on the bridge, and in a coalition of this type you are constantly learning. In fact, the most challenging thing about my role has probably been maintaining continuity when people move on – the turnover of people on whom we depend to make things happen is tough. The hard-core group of us who have been involved long-term is quite small. I rely a lot on my deputy, Mark Watson, who has worked in heritage longer than I have. He's an encyclopaedia on legs! Having been through the UNESCO process with New Lanark, he was invaluable! Best of all, his brain works differently from mine, which I really appreciate.'

Leith-based Dr Oglethorpe now leads the life of a jetsetter. 'I never sought global exposure, really, but that is how it's turned out. I am now the president of an organisation called The International Committee for the Conservation of Industrial Heritage, TICCIH for short, and in many ways the Forth Bridge has been my ambassador. It started in 2000 when we took 100 delegates up to the top. Of course, there are many international connections to the bridge's heritage, particularly to Japan where Forth Bridge engineer Watanabe is a respected key figure. It's part of my job to write papers for all sorts of publications and promote our heritage, particularly the Forth Bridge, on a global stage.'

One of the undisputed highs of Oglethorpe's tenure was the Forth Bridge inscription on UNESCO's World Heritage list. A last-minute nomination in 2011 from Fife Council ensured the structure's inclusion on the UK's updated Tentative List, the first step on the road to World Heritage status. Oglethorpe recalls the frustration at the time: 'Historic Scotland wasn't allowed to nominate any site – someone else had to do it. Thank goodness for Fife Council! It felt appropriate as the Forth Bridge forms part of their logo.' After a rigorous process to ensure that all the criteria were met, the Forth Bridge finally received the honour. Oglethorpe recalls the time fondly: 'We travelled to Bonn – it was the 5th of July 2015, and a very special experience. The nomination was put forward and the committee then deliberated, but it was actually proposed live in French, much to our delight – the Auld Alliance and all that! In the end, the nomination was commended for its quality, and for the fact that it was so lean and succinct – the dossier weighed only 889g. Many nominations are enormous and therefore take a huge amount of work to assess.'

Miles Oglethorpe clearly gets a buzz from preserving what might otherwise have been lost. 'One day back in 1987, I got a phone call from my former boss: "Miles, they are closing Arrol's Dalmarnock Ironworks. People are pilfering, get down there quick!" William Arrol's was the engineering company which built the Forth Bridge, and many other famous structures too. We rescued the collection – if you can believe it, the company records were lying on the floor with boot prints all over them. We spent years cataloguing it all. This factory fabricated metal structures for buildings and bridges all over the world, including Tower Bridge in London!'

As a global ambassador for the Forth Bridge, Oglethorpe relies heavily on powerful images to tell the Forth Bridge story. However, taking a good photograph of the bridge is not as easy as one might expect, given its extraordinary geometry. Oglethorpe reflects: 'It's very hard to create an image of the bridge as a whole and do it justice. You have to approach it with purpose and ask yourself: Which of its extraordinary qualities are you wanting to capture? For me, it's the amazing geometry of the steelwork that stays in the memory.'

VOLUNTEER
Len Saunders

'MY FAVOURITE VIEW of the Forth Bridge? Sitting in a rowing boat, passing through between the legs of the bridge, with the width of the boat and oars just fitting the space. You stop rowing and look up. And then it's just a cathedral of steel! It's a view most people don't get, an amazing sight. It has an effect on you.'

A member of the South Queensferry Rowing Club, the retired engineer Len Saunders has become something of a Forth Bridge enthusiast. Perhaps unusually, he remembers exactly how his interest in the structure was first sparked: 'When I was around 11, I was given a book called *Victories of the Engineer* by A. Williams. It had a chapter on the construction of the Forth Bridge, and an image of a steam train crossing it on the cover which has stuck with me ever since.'

Raised in Edinburgh's Davidson's Mains, his parents took Len to see the Forth Road Bridge being built. 'We also often crossed on the ferry to visit relatives in Fife, leaving from the Hawes Pier,' he recalls. 'I remember seeing the Forth Bridge from the ferry. Little did I know then that I would live in Queensferry for most of my life. But that's where we settled when we got married – it's a nice place to live.'

Len became interested in local history in the late 1980s and joined the South Queensferry History Group. However, he stayed clear of seeking out what he calls *Forth Bridge stuff*. 'I thought, everybody already knows so much about it.' That is, until James Walker another member of the History Group suggested preparing some illustrated talks on how the Forth Bridge was built – Len joined forces with him. Together, they took their Forth Bridge talks across Edinburgh. When James dropped out, Len collaborated with Frank Hay – also a History Group member – and with a range of updated presentations the pair have now delivered 90 plus talks *from Hawick to Helensburgh*, as they like to say.

'People are interested,' Len states. 'There is just something about that bridge that gets in people's psyche. For a start, there are countless legends and anecdotes, and it's sometimes hard to discern fact from fiction. For example, there is an urban myth of two workers being trapped in the caissons, after accidentally being

concreted in. The legend alleges that the pair of trapped workers were supplied with poisoned food to hasten their deaths and ease their passing. There is no evidence that this actually happened.'

Like Frank Hay and Gordon Muir, Len was involved in the research group identifying the names of the fallen briggers for the memorials. He recalls: 'It was a rigorous process – we needed to be certain that these people died on the bridge during construction, or as a direct result of injuries sustained there. We needed evidence. But we also needed money if we wanted these memorials built. To start with, we approached all the local businesses which feature the bridge in their various logos, and we also asked big companies including Network Rail, Balfour Beatty, Ove Arup and AG Barr, the makers of Irn-Bru. Many donations were also received from individuals who wanted to contribute.'

The unveiling finally took place on 18 May 2012. Len describes the scene for me: 'It was a wild, horrible day, with the *Maid of the Forth* taking the participants on board and crossing from North Queensferry to the Hawes Pier on the south side. My rowing club friends rowed across behind the tour boat to take part in the day, and I remember worrying about them – I could tell that they were struggling in the challenging conditions! I have since roped quite a few of my rowing club and other friends into Forth Bridge-related things. Some of them helped us with talks at the last Barnardo's fundraising event where we explained the history of the bridge and the riveting process. And some also help me to maintain the memorials.'

The two memorials can be susceptible to the effects of being so close to the sea. 'You need to wash the salt off them and rub them with Brasso – then they need a wax and polish. When my rowing friends come along, I always warn them in advance that while we are cleaning the memorials, someone is bound to come up and start a conversation about the briggers, and so it happens – every time! Even when I am passing at a distance and see someone looking at the inscriptions, I can't help myself – I have to go and speak to them, to explain. I get a kick out of it, I really do, because whenever I look at the memorial, I remember the stories. Each name has a story behind it! And yes, I'm proud of being part of that journey. Even when the money wasn't coming together and things looked bleak, the committee never gave up; we met every month for over seven years. It was such an important project. If you keep going, you have a chance!'

Working together so closely forges connections. On one occasion, he accompanied his fellow Forth Bridge researcher Jenni Meldrum, an antiques shop owner, to nearby stately home Hopetoun House for the filming of an episode of the BBC's *Antiques Roadshow*. 'I was really only there for moral support. It was a sunny day, and it was later reported that the day's audience was the biggest the programme had ever attracted, more than 4,000 people. The queue was huge. After some filming we were eating a picnic in the temporary car park, and we got chatting with an old man who we discovered was carrying a bag full of Forth Bridge rivets! It turned out that his grandfather had worked on the construction of the bridge, and best of all, he had photographs of his grandfather. It was then less than a week until *The Briggers* book was set to go to print, but an image of this man as a young boy holding hands with his grandfather made it into the book. If you talk to the people you bump into, you can discover amazing stories.'

With a mechanical engineering background, Len spent a lot of his working life as a Chartered Health and Safety Manager. 'It is amazing to see the differences between how things were done then and the multitude of precautions in place nowadays to protect workers on the bridge. "Health and Safety" is often ridiculed. But when you see the conditions the briggers were working in and the price many of them paid – Health and Safety is no joke. It's an absolute necessity!'

Much of Len's time is now devoted to the Forth Bridge in one way or another. 'It's funny how all of this started for me with reading a book as a kid,' he reflects. 'The Forth Bridge has now become a really big thing in my life. Huge, in fact.'

WRITER
Elspeth Wills

'FRANK, ONE OF the volunteers, arrived and dropped three huge boxes of research on my study floor,' the writer Elspeth Wills tells me. 'Glutton for punishment that I am, I read through it all. That's when I realised how much of the story they had missed! I knew then that I wanted the job of writing this book.'

Elspeth was the ideal candidate to write *The Briggers: The Story of the Men Who Built the Forth Bridge.* Not only was she an experienced and respected non-fiction writer with a long list of publication credits, but she had also made a significant contribution to South Queensferry by researching and writing the content for local signage, working in collaboration with designer Gordon Muir and landscape architect Paul Hogarth. In addition, she had updated a book of local walks and contributed to a range of publications and pamphlets aimed at tourists on both shores of the Forth.

However, the truth is that the job of writing about men who built the Forth Bridge came at the right time for her, too. Her husband Michael had just died. 'He was my sternest critic and I missed him terribly,' she explains. Their partnership extended beyond the personal, co-writing and working together often. It was little wonder that Elspeth was drawn to this collaborative project which appealed to her interest in local and social history. A handful of South Queensferry-based history enthusiasts had identified the names of the fallen briggers for the memorials. The task now was to turn the mountains of research into a book. Elspeth remembers: 'A genealogist had helped the group amass an immense amount of detail, and as it stood, it was a book about the dead. "A book about the dead? That's never going to sell!" I told them. But I did lower my usual rate because I really wanted the job – the subject personally fired me up.'

As part of the assignment, Elspeth had to work closely with the team of researchers, a process she relished. 'The project was a lifesaver for me at the time. I used to meet the others every month at the Hawes Inn which was of course already trading at the time of the Forth Bridge construction. Jenni and I were the only women and mostly interested in the social history aspect. We would have supper together first and then the men arrived. They were all engineers and

much more focused on the technology and innovation angle. Once the structure of the book was agreed, I showed them a chapter a month, and we would discuss it all. They were very easy to work with. But I did a little research myself too. One of the Forth Bridge assistant engineers, Wilhelm Westhofen, documented the process in detail and it is seen as a bit of an engineering bible – but I found his writing impenetrable. Hands up anyone who knows how the Forth Bridge used skewbacks? No, I preferred a light touch when it came to technology and such. However, I became fascinated by a newspaper which an enterprising individual put together during the period of the Forth Bridge construction. There were all these fascinating stories of injuries – and of drink, many of those!'

At the time, the group hoped to source local government funding, but Elspeth had other ideas. 'I knew from bitter experience with funding applications that you need to prove that you cannot raise the money commercially. So, I suggested trying a few publishers first. Their rejections would strengthen our case. To our astonishment, we received two offers of publication, and *The Briggers* book went through a traditional publishing process. None of us could have foreseen that!'

Despite her extensive knowledge of the Forth Bridge and its people past and present, Elspeth has never ventured to the top. 'Let me put it this way: I'd love, in THEORY, to do it. But no, I have no wish to!' she says emphatically.

Elspeth is old enough to remember the ferries as a child and recalls hearing stories of American troops stationed at North Queensferry throwing dollars from the train windows for luck, instead of the traditional pennies – much to the delight of the local population. For half a century, she has lived in her flat in central Edinburgh, overlooking the castle. She reflects: 'Edinburgh and South Queensferry have changed hugely over the centuries. But yes, I do feel an affinity with the area around the bridge. Somehow, North Queensferry never seems to be as much written about, but it has a lot of interesting Forth Bridge stories of its own.'

Elspeth strikes me as warm, forthright, and above all curious. I am deeply impressed by her writing which touches on such a wide range of subjects, from cruise ships to football badges, innovations, local and national history, place names, walks, tourism and much in between. Her current work-in-progress, she tells me, explores the history of Britain through pub signs. If I am honest, I feel a little intimidated by writing up the interview and letting her see it. When I hint at this, Elspeth laughs.

'If someone tells me: Elspeth, write a book about X, or Y, or Z, I will gladly do it. But believe me, I wouldn't have the first idea of what to write about myself. I will leave that to you.'

X-MAN
Donald Scott

'IT WAS AN April morning. When you are working on the bridge, you don't say that you're feeling dizzy. You're going nowhere if you say that. I *was* feeling dizzy that day, but I said nothing. But when I popped out of the wee lunchroom at track level because I needed some air, I just collapsed. After that I remember nothing, other than waking up in the hospital at Kirkcaldy with tubes coming out of me and x-rays being taken.' Donald Scott had suffered a cardiac arrest on the Forth Bridge.

He had recently returned to Scotland after 30 years of working in television production in Canada. Originally a BBC cameraman, Donald and his wife moved overseas to join his brother who had emigrated. By the time the couple returned, Donald was almost 60 years old. His nephew suggested trying for a job on the bridge for its refurbishment. 'So, I said aye,' he says matter-of-factly.

I am astonished. How did he cope with the physical challenge? He chuckles. 'I lost a lot of weight, put it that way. We used to have a nickname for the bridge: Big Red.'

'The initial induction to the job took place on shore, but then they take you up,' he reminisces. 'And they ask if you're ok with heights. Some of the new starts would arrive, say "I can't do this" and walk off again. I remember going up. It was 7.30am in November and dark, and we were told: If you don't have sight and sound of the safety ship, they can't see or hear you either – that was a ship that patrolled beneath where we were working. Your overalls had these tags attached to them. God forbid, but if someone should fall, they may only be able to identify you by the tags.'

Donald worked as a sheeter, a job that took him to the structure's most precarious locations. Teams of eight to ten men had to manually take rolls of thick plastic sheeting up the bridge from track level – no hoist was big enough for these loads. They then attached the sheeting to the outside of the scaffolding to allow the refurbishment work to get underway. 'It was windy, like. But the health and safety was impeccable, I'll say that. Wherever we were working, the foreman was going there first, checking everything was all right. The scariest thing was swinging round to the outside where one of your colleagues hands you the heat gun to shrink

the sheeting into place. Some guys couldn't do it. I took my turn though – and I couldn't believe how quiet it was. Thank God I wasn't doing that on the day I took unwell though. If I had been on a platform instead of at track level, they would never have got me down in time. I'd be gone.'

As it happened, Scott was lowered to the safety boat by crane and rushed to shore. By an extraordinary turn of good fortune, two paramedics were having coffee nearby and hurried to help. 'My nephew came off the bridge as fast as he could to see how I was, and he told me afterwards that he saw one of them shaking his head to the other. They didn't think I was going to make it. It was close. I was 63. They had to use shears to cut off my clothing so I could get treatment quickly.'

Donald narrates all this dispassionately, as if the dramatic incident had little to do with him. But then he is the type of man to seek out a physical construction job at great height in his sixties, while also working as a cleaner for a landscape architecture firm. 'I was just driving past the Paul Hogarth company on Hawes Brae, and I saw an advert for a cleaner. I popped in for a chat and stayed for 17 years,' he laughs.

Now fully retired, Donald is humble. 'I'm not sure I can be really that much use to you,' he says, completely unaware that interviews like this are gold dust – here is a man who can lift the curtain on what really went on at the Forth Bridge. He recalls: 'Once there was a bunch of firemen who came to see what to do in the event of a fire. We entered with them from the Hawes walkway – you can walk from one end to the other from there. One of them got white knuckles and froze. It took four of our guys to take him off the bridge again. Four! But actually, that's how the experienced workers sometimes got their entertainment. They'd deliberately tell the new starts to keep looking at the water, as if that was the best way to avoid vertigo. In fact, looking at the water is the worst thing you can do – that's the way you get dizzy. A bit cruel I suppose.' He can't resist a chuckle himself.

'I can think of many memorable moments though. One day they told us that the *Flying Scotsman* was coming across the bridge.

When the time came, we all stood at the trackside and watched it, that was nice. Another time I was down at the caissons and noticed indentations, as if they had been hit by golf balls. I learned then that these marks were created by German bullets bouncing off, isn't that something? And there was another time that the local MP Tam Dalyell visited. He told me the story that during the Second World War, Josef Goebbels used a picture of the bridge under construction and claimed to have destroyed it for propaganda.'

Tales of dramatic rescues and daring endeavours – I can picture these tense scenarios, but somehow, the one which lodges itself into my memory is a gentler tale. Birds had built a nest in an area where Donald and his team of sheeters were going to work next. He tells me: 'These are tough guys, like, the ones who work on the bridge. But there were chicks in there, and our foreman told us: "Don't sheet that area until the birds leave."'

YOUR VIEW ORGANISER
Jordyn Armstrong

'IT WAS MEANT to be a one-off. That's honestly what I thought.'

Jordyn Armstrong is young, energetic and cheerful, and brimming with enthusiasm for the event she helped bring into the world.

She tells me how it all began. 'I worked for the charity Barnardo's at the time, and we were exploring a partnership with Network Rail. My colleague and I came out of the meeting where we had discussed all the usual stuff: bake sales, fun runs and so on, when we were introduced to Colin Hardie of Balfour Beatty. To be honest, this whole event was his brainchild. He just said: "You know, we have a platform up there, and there is a hoist – we could perhaps sell tickets to take people up, as a once-in-a-lifetime opportunity." We turned and went right back into the room to chat about that idea. And what a journey it has been since then.' Jordyn smiles, but with a little nostalgia too – Your View used to dominate her working life until she left the charity in 2021.

What were these early days like? She laughs. 'I have to admit, I was so naïve. I grew up in Dumfries and Galloway, so I had no idea what the bridge meant to people, its history, and its role in the community. All I knew was that we were excited about the prospect of running the event. One Friday in summer, I priced the tickets at £50, set up the Eventbrite link, shared it with the Balfour Beatty and Network Rail guys and signed off. I was attending a wedding the following day, so I was out of action, but I returned to a bunch of missed calls from the Supporter Relations Team: "Your website has crashed, Jordyn, it's not working. We've had hundreds of calls all weekend." I was horrified.' Jordyn throws her hands in the air as she relives her panic: 'I was so stressed, and probably still a bit hungover from the wedding, but I rushed to my laptop. It turned out that the booking page hadn't crashed at all – it had just sold out – in SECONDS.' She shakes her head, as if surprised all over again. 'There were so many tweets and Facebook shares – I honestly couldn't believe it! I'd been involved in organising hundreds of events and you never sell all the tickets, never!'

There was only one thing to do. She contacted her partners at Network Rail and Balfour Beatty and asked if they would consider doubling up and running the event for an additional day. They agreed, but those tickets, too, sold in a flash. 'That's when it hit

home, I think. This was going to be a really good event! I must admit, at the time the bridge was just a bridge to me, and the event was just a job. But over the years, the bridge and the Your View event became a genuine passion.'

The first year was a simple set-up: groups were limited to 13 people at a time, constrained by the size of the hoist at the North Queensferry site. There was no catering, no merchandise, no extras, just a couple of marquees. After that unexpected demand, the team doubled capacity again for 2018, running the event for four days over two weekends. By 2019, Your View had transformed into a six-day event, with input from local community groups, sunrise and sunset packages from dawn to dusk, merchandise and catering. Jordyn tells me: 'We just didn't stop. Over the course we took 3,600 people up the bridge and raised almost a quarter of a million pounds for the charity. It's mind-blowing when you think about it. We had 200 volunteers, from Ikea, the banks, plus our local partnerships too – it was very important to us to keep the locals on side.'

Her excitement at the memories is palpable and I feel energised just listening. She tells me that the team were on site from 5am, before daybreak. The first group would ascend in the hoist in time for sunrise. It is clear that what matters to Jordyn, first and foremost, is meeting people – and she remembers many of them clearly.

'We made sure there was time to speak to people personally. I'd begin with a welcome talk, then the Briggers would give their demonstration, the safety briefing was next and then it was off to the top. But I loved finding out where the individuals came from, and their reasons for coming. It's called Your View after all – everyone has their own perspective. One husband had given the trip to his wife as a secret wedding gift; it was hilarious: "Am I bungee jumping?" she asked – she genuinely had no idea why she was there! Of course, the husband was winding her up all the way, fuelling the fire. Yes, we did have a few who were only there for that Instagram shot, but they were a tiny minority. Most had reasons for being there: family trips, bridge enthusiasts, visitors from all over the UK and beyond.'

YOUR VIEW ORGANISER: JORDYN ARMSTRONG

Jordyn's colleague Kat gathered the stories – one visitor turned out to be the grand-niece of Sir William Arrol, the engineer responsible for building the Forth Bridge. Someone else played the bagpipes on top of the North Cantilever platform, and another couple got engaged – even bringing plastic champagne flutes and bubbly so that the whole group could join them in the celebrations. A French visitor had planned his whole trip around the Your View experience while a couple from Texas had flown across the pond for a long weekend just to attend. Jordyn's smile gets wider and wider as she hits her stride: 'In 2019, the last year I ran the event, the HMS *Prince of Wales* aircraft carrier went under the bridge; it was incredible. And once a group of swimmers crossed the Forth – they hadn't known the event was on and didn't have tickets. We managed to squeeze them in for an extra run to the top which was so lovely.'

The team encountered some challenges too: 'We had a daughter who had bought the tickets as a gift for her dad, but as soon as she got in the hoist, she was just terrified. She wanted to get down, but the whole group would have missed out then. The volunteers were so good with her, reassuring her and comforting her all the way, and she made it. She was so grateful and said that she couldn't have done it without us. Colin and his team were amazing, giving so generously of their own time. It was physically exhausting, but we were all one big team – it felt like a family affair. Even my mum joined in, writing the certificates out by hand!'

Has she been back to the event since she left the charity?

She looks a little wistful. 'When I left, towards the end of the pandemic, we weren't sure if Your View would ever happen again. It had been my baby for three years; three very successful years. If I had known it would resume, I probably would have struggled to walk away. Having been so heavily involved in the past, I think I might find it difficult just to watch! But I feel so grateful for my part in it, and so proud of everything we achieved. Your View was a finalist for the Third Sector Awards in 2019 and the feedback over the years has been incredible. Without doubt, it is one of my biggest career achievements, but more than that, it's about the people you meet and the connections you forge.'

She nods for emphasis. 'No job will ever be as rewarding as that Barnardo's job for me – and it was all down to Your View.'

ZOOLOGIST
Cristín Lambert

MADE FROM GIRDERS: OUR FORTH BRIDGE

'I ACTUALLY WAKE up to the Forth Bridge every day,' Cristín Lambert tells me with a laugh. 'I have an AI print of a painting of it in my house. I don't even know why I have it – I guess it's just iconic. Every time I cross over on the train, I have to take a photo.' The young Irishwoman works as a conservation officer for the Fife Coast and Countryside Trust where she covers the area of West Fife – including the Firth of Forth shoreline around the Forth Bridge.

My zoologist is a slightly hesitant interviewee at first: 'I definitely have a bit of imposter syndrome going here. Even though I have lived in Fife for four years, I am still relatively new to this job,' she admits. I assure her that she is exactly what I need. After all, I seek to profile the people, the communities around the Forth Bridge, and newcomers are as welcome as those who have known the bridge for decades. She belongs, just like everyone else.

Aged only 18, Cristín left her native Dundalk in Ireland to study Zoology at the University of Aberdeen. 'I thought it would be a four-year thing and I'd be back, but that's not how it turned out,' she explains. Instead, she secured work in Scotland, including a conservation internship as a seasonal ranger on the Isle of May. 'That was my first introduction to the Firth of Forth. It was a really varied role: visitor tours, seabird counts, ringing seabirds and monitoring populations. The island is among the best for historical research in that area.' Following a Master's degree in animal behaviour from St Andrews University, she worked on beaver reintroduction for NatureScot before becoming the Research and Data Manager for the John Muir Trust. However, something didn't sit right with her. 'I was in a desk-based role, managing data. But I didn't study zoology to be sitting at a desk all day! The thing is, once you move from a desk-based job back outdoors again, there is usually a pay cut. You definitely don't do it for the money!' she reflects. 'If you work in policy, you may make a difference, but you don't always get to see tangible changes. In this sort of job that I have now where you do the smaller scale stuff, you do see tangible progress: improved biodiversity on that stretch of coast you've been working on, for example. And that gives me real satisfaction.'

While the Firth of Forth coastline around the bridges is part of

'her patch', Cristín may also be involved in improving the Lyne Burn in Dunfermline or out in the Lomond Hills with sweeping views of the estuary. What excites her most about her job, I want to know.

'I love all animals and find studying nature completely fascinating. As a zoologist I enjoy survey work – surveying wetland birds, breeding bird surveys, biodiversity and so on. But my current remit also includes community engagement here in West Fife. At the moment I am planning a conservation engagement programme which includes activities for a wide range of people. It's really about communicating to the community what they can do for their natural heritage. Think tree planting, wildflower meadows, bird identification and all of that.'

Cristín's commitment and passion for the natural environment is palpable. What changes would she like to see to protect it better?

'In terms of marine protection around our shores, I feel as though we are lagging behind a bit,' she answers after a moment's thought. 'Scotland has so much coastline, and the breadth of marine life is just ridiculous, world-class. For example, we have the biggest colony of Northern gannets in the world right here in the Firth of Forth on Bass Rock. Yes, we can watch nature programmes on the telly and marvel at huge herds of animals in Africa – but what about the 200,000 charismatic seabirds we have right off our shore on the Isle of May: razorbills, puffins, fulmars and guillemots and so many others. The Forth supports internationally important numbers of waterfowl like knot, eider and bar-tailed godwit. Not to mention the marine mammals like seals, whales and dolphins which can sometimes be spotted from the bridge! Our marine wildlife here in Scotland is spectacular, unbelievable – it's as good as any around the world! It's really encouraging to see projects like Restoration Forth taking place to restore seagrass habitats and native oyster populations in the Firth of Forth. I'd love to see stronger marine protections put in place to protect and enhance our marine biodiversity in the face of the climate and biodiversity crisis.'

Postscript

THE LITERAL-MINDED technical puritans may have an issue with my slightly cheeky interpretation of the phrase *Made from Girders*. Wasn't that misleading? As Writer-in-Residence, wasn't I supposed to write about steel and concrete? Shouldn't I restrict myself to the bridge, and the bridge only?

My simple defence is that I am an artsy type. In that sense, I love the Forth Bridge as a metaphor, not just as a physical structure. For me, the Forth Bridge spans more than a body of water – it spans centuries, communities, land and sea, past and present. Just like the six-and-a-half million rivets holding it together, there are countless points of contact between people and places all around the bridge. Engineers and environmentalists, managers and mental health activists, construction workers, cleaners and campaigners. We need each other, rely on each other, rub up against each other, pass each other by – but ultimately, our communities resemble the majestic Forth Bridge in their midst.

I had tremendous fun speaking to all the individuals featured amid these pages. People are so interesting! I had the privilege of meeting some in person while others chatted to me on the phone or via video. I felt a connection with them all.

If only the alphabetical chapters had been as straightforward! You would not believe the amount of re-jigging and re-labelling required. And I won't apologise for some of the more contrived chapter headings (Lovebirds? X-Man? I know, I know…) because those proved a particular source of amusement to me throughout the writing process. The beginning was easy – all the letters of the alphabet were available. It became much more challenging when only the tricky ones remained. Praise be for Queensferry! Thanks to the place name, the options for Q were almost unlimited, and I chose to cheat by including two contributors instead of one.

My promise to those trusting me with their stories was that they would have a chance to correct or if necessary, veto my drafts. As a result, some chapters went through extensive drafts and consultations while others accepted my write-ups with enthusiasm. I was grateful for both.

However, one sentiment united almost every person I spoke to – and on reflection, after a year as Writer-in-Residence here I realised that I share it too.

They all claimed the bridge as their own.

Barbara Henderson

Luath Press Limited

committed to publishing well written books worth reading

LUATH PRESS takes its name from Robert Burns, whose little collie Luath (*Gael.*, swift or nimble) tripped up Jean Armour at a wedding and gave him the chance to speak to the woman who was to be his wife and the abiding love of his life. Burns called one of the 'Twa Dogs' Luath after Cuchullin's hunting dog in Ossian's *Fingal*. Luath Press was established in 1981 in the heart of Burns country, and is now based a few steps up the road from Burns' first lodgings on Edinburgh's Royal Mile. Luath offers you distinctive writing with a hint of unexpected pleasures.
Most bookshops in the UK, the US, Canada, Australia, New Zealand and parts of Europe, either carry our books in stock or can order them for you. To order direct from us, please send a £sterling cheque, postal order, international money order or your credit card details (number, address of cardholder and expiry date) to us at the address below. Please add post and packing as follows: UK – £1.00 per delivery address; overseas surface mail – £2.50 per delivery address; overseas airmail – £3.50 for the first book to each delivery address, plus £1.00 for each additional book by airmail to the same address. If your order is a gift, we will happily enclose your card or message at no extra charge.

Luath Press Limited
543/2 Castlehill
The Royal Mile
Edinburgh EH1 2ND
Scotland
Telephone: 0131 225 4326 (24 hours)
Email: sales@luath.co.uk
Website: www.luath.co.uk